ECOLOGY AND ENVIRONMENT

ECOLOGY AND ENVIRONMENT

R N BHARGAVA • V RAJARAM • KEITH OLSON • LYNN TIEDE

CRC Press
Taylor & Francis Group
Boca Raton London New York

CRC Press is an imprint of the
Taylor & Francis Group, an **informa** business

The Energy and Resources Institute

CRC Press
Taylor & Francis Group
6000 Broken Sound Parkway NW, Suite 300
Boca Raton, FL 33487-2742

First issued in paperback 2023

ISBN 13: 978-1-03-265386-0 (pbk)
ISBN 13: 978-0-367-17882-6 (hbk)
ISBN 13: 978-0-429-05822-6 (ebk)

DOI: 10.1201/9780429058226

Print edition not for sale in South Asia (India, Sri Lanka, Nepal, Bangladesh, Pakistan or
Bhutan)

Library of Congress Cataloging in Publication Data
A catalog record has been requested

Visit the Taylor & Francis Web site at
http://www.taylorandfrancis.com

and the CRC Press Web site at
http://www.crcpress.com

Foreword

"In the race for development, which ideally ought to improve the quality of life of the citizen, the relationship with environment is often lost sight of. That environment is sacrosanct; that the knowledge and application of science warrants harmonious use of natural resources without destabilizing the cycles of nature; that the purity of air, water and land has been inherited by a generation in mortgage for children of tomorrow; that it is implicitly imperative for each generation to leave the environment to the next generation in a better state than they found it."

I give excerpts from my book 'India at Turning Point' below: "On the drive from Tiruchy airport to Karur that afternoon in 2013, I was quite surprised to see that Kaveri River, which always had plenty of water, one of the perennial rivers mentioned in our scriptures, appeared to be completely dry."... "As I reached Karur, drove close to Amaravati, I noted that there was not a single drop of water in the river. As I walked upto the canal alongside the temple, I was astonished to see not a drop of water in it; in fact it had been converted into a sewer canal, where all the wastewater and sewerage was diverted – uncovered as it was. The canal – surely it has to be called sewerage ditch now – raised a stink as I approached it. In the course of one day, I had discovered that three water channels, the Kaveri River, the Amaravati River, and the village canal, all of which had 24X7X365 water flow had all become totally and probably irrevocably dry – indeed the canal had become an effluent and sewer discharge passage."... "Next day, it was quite dark as I crossed the Yamuna by the DND on my way home; suddenly a stench hit my nostrils – I asked the driver what it was; he looked surprised, mentioned to me that it was the foul smell emanating from the Yamuna – yes indeed even the Yamuna has now become an effluent discharge canal. So much for our perennial rivers, mentioned in daily prayers: गंगे च यमुने चैव गोदावरी सरस्वती, नर्मदे सिन्धु कावेरी ... ... referring to the perennial rivers of India – Ganga, Yamuna, Godavari, Saraswati, Narmada, Sindhu and Kaveri; Saraswati has long since disappeared underground, alas literally and figuratively; Sindhu (Indus) has been ceded to Pakistan;

in one single day I discovered that Kaveri had become totally dry, and Yamuna is now a sewage canal; the count is on – it is a question of time before the others also go the same way. Within a month or so thereafter, one learnt that the Mandakini, a tributary of the Ganga got flooded for a brief period, caused untold destruction and desolation, and swiftly ran on to the Bay of Bengal, to return to its normal thin stream. I wonder if there is any policy recognition of the criticality of the situation, the impending catastrophe – what the future has to portend in terms of water supply in India. A few years back, one heard of a grandiose scheme to connect the major rivers of India – I do not know what the feasibility is. However, the way things are moving now, it may actually amount to creating a national sewage grid – I cannot comment on its desirability... A country's wealth is measured by the amount of water it has.".... "The water table in Noida is visibly going down; Gurgaon has nearly reached panic stage with regard to water; in Bangalore, a cruel water-mafia is emerging, supplying potable water to colonies at robbery rates; the recent drought in Maharashtra has an entire state reeling. Is the looming crisis seen by anybody?"

In recent years, Delhi has acquired fame for having the worst air pollution in the world; Mumbai is not far behind – many other Indian cities already figure in the international roll of honour in this regard. Reckless expansion of vehicular traffic, uncontrolled particulate emission levels and sheer mismanagement, indeed lack of any meaningful management, has brought us to this level.

The High Level Committee (HLC) established by the Ministry of Environment and Forests (2014) has, inter alia, found that "The primary focus of environmental and forest governance in the country needs to be re-aligned through a series of structural and process-oriented changes. While the pace of diversion of forest land has decreased in recent years, the target of 33% of land area as forest cover is a long way off; the more disturbing aspect is that the quality of forest cover has seen a secular decline. New forestation policies to attract investment of growing forests in private land, is imperative."

Sustainability is the key word. While the imperatives of development cannot be overlooked in a country where over 75% of the population is poor, it is even more important that environmental considerations should be given the highest priority. The issue should not be seen as man vs. animal, or development vs. climate; technology should be appropriately used to optimally manage maximum benefits, without loss to ecology considerations.

Bharat has always been conscious of the supreme position given to nature. The Vedas and Upanishads are replete with chants and ideas strongly supportive of the respect given to nature, and the need to preserve it. If this was so, when the population density was a miniscule fraction of the current position, this concern and regard of ecology has to take a dominant position in our national planning and development. In the governance model for the first few decades of after independence, the issue of environment or climate never figured as a serious factor. It is indeed only in the past couple of decades that the imperative urgency of factoring in environmental consideration in every aspect of national life has acquired renewed status. One can freely predict that issues relating to environment and climate change will be foremost in all issues relating to national development and project planning in the future.

Issues relating to municipal solid waste management, control of vehicular air pollution, management of efficient discharge from commercial and industrial establishments, and optimal management of industrial air and other pollutions have now reached centre-stage. These issues have been tackled in other parts of the world and many examples are available for adoption with suitable modifications in India. The internationally accepted Principles of 'Polluter Pays', and the legal doctrines of 'proportionality', 'sustainable development' and 'inter-generational equity', the 'doctrine of marginal appreciation' have all been the basis for evolution of policy in this country; much has to be due in bringing these into practical application.

"Universal human dependence on the use of environmental resources for the most basic needs renders it impossible to refrain from altering environment......environmental conflicts are ineradicable and environmental protection is always a matter of degree, inescapably requiring choices as to the appropriate level of environmental protection and the risks which are to be regulated. This aspect is recognized by the concept of 'sustainable development'. Setting the standards of environmental protection involves mediating conflicting visions of what is of value in human life" – Chief Justice Kapadia in 'Lafarge' case.

Much of the available literature on the subject relates to analysis of judicial pronouncements in the field, or polemical statements by 'environment lobbies' expressing extreme views. The HLC has recommended a 'Green Awareness' programme, including interweaving issues relating to environment, in the primary and secondary school curriculum. In this context, this publication *Ecology and Environment* (Global Perspective) written by R N Bhargava, Raj Rajaram, Keith Olson, and Lynn Tiede is a welcome addition to the literature available to the public in this field. In a sense, it is also a path-breaking effort to describe, in simple layman language, the broad issues involved with balanced analysis. The time has now come for every child and citizen of the country to become aware of ecology issues – this book is a welcome new arrival for this important purpose.

The authors have also added a chapter on global climate issues. This is one of the foremost themes engaging mankind as a whole, that in the race of development and competition, we should not damage the natural balances; great care is required to guide our efforts for the future. This is one of the major challenges facing the global leaders today, how in a competitive atmosphere, the common theme of maintenance of global ecology will not be relegated to lesser importance.

I am happy to note the arrival of this book, which one hopes will get much currency and usage in the country, and will benefit particularly the younger generation.

T S R Subramanian
IAS (Retd)
Ex-Chief Secretary
Government of Uttar Pradesh
Ex-Cabinet Secretary
Goverment of India

Preface

The alarming rise in greenhouse gas and pollution level which has resulted in serious environmental and ecological harm is the biggest concern today. It has not only made the lives of mankind miserable but also threatens their very existence. The survival of human beings on earth depends on the availability of clean environment. The nature can be protected only when there is a good understanding of the ecology. For this, young people and the general public have to become aware of the underlying issues associated with maintaining a clean environment and protecting the nature.

Ecology and Environment has delved in depth on the subject and brings a broad perspective of the various issues and how readers can get actively involved in creating an environment and ecology that promotes health and biodiversity. Following the curriculum of University Grants Commission, the book begins with a description of sustainable ecosystems. Case studies from around the world have been provided. The book discusses in detail our natural resources, both renewable and non-renewable. Forests, minerals, water, and land resources are also covered. Renewable energy sources—solar, wind, hydropower, and biomass energy—are described in detail. Non-renewable sources, such as coal, oil and gas and nuclear energy, and their impact on climate change are extensively dealt with in the book.

The chapters on biodiversity and population explain how the population explosion is impacting our planet. A complete chapter is devoted to pollution and its impact on human health and the environment. An important aspect of this book is the detailed discussion on social issues associated with maintaining a clean environment and how public participation can ensure that we work constantly towards it. The science of climate change and what we can do to mitigate its impacts is also explained.

India has a rich and long framework for protecting its environment and ecology. It is the only country that has included climate change as part of the Ministry of Environment and Forests. The environmental legislative framework established by the Indian Parliament and the agencies that implement this framework are also discussed.

The book will provide guidance and knowledge to students and those interested in clean environment and protection of ecology for a sustainable future. The readers, with this broad background, will learn how they can advocate for a sustainable and clean environment.

Mr R N Bhargava would like to dedicate this book to the youth of the country who inspired him to write this book. He would also like to thank Shri T S R Subramanian for his guidance. Dr Rajaram would like to dedicate this book to his wife Usha Rajaram. He would like to thank his daughter Pooja Rajaram and Ms Jennifer Freeman for their editing assistance. Mr Keith Olson dedicates the book to the school children and the environmentalists he has known since 1977 who work towards a livable world community. Ms Lynn Tiede dedicates the book to the people with whom she interacted regarding environmental facilities and education initiatives during her visit in India in 2005, China in 2009, and South Africa in 2012.

R N Bhargava
V Rajaram
Keith Olson
Lynn Tiede

Contents

1
Multidisciplinary Nature of Environmental Studies

1.1 HUMANS AND THE ENVIRONMENT

'Environment, being green, live green, go green, and sustainability' are the catch phrases used in the present day. However, the concepts are not new. Societies of the past and many populations, who still live in rural settings, as is the case with 70% of India's population, seem to intuitively have an understanding of the delicate dynamics of the environment and ecosystem. Those who live in poverty, again about 70% of India's population, respect and have an innate sense of the importance of conserving natural resources and create smaller carbon footprints (the measure of a person's total lifestyle demand on nature). Hence, while the discussion and greater awareness of the environment in the modern age by urban dwellers is relatively new; human dependency upon the environment and ecosystems, and the need to live in partnership with it, is not new at all.

The environment is defined as surroundings. This includes natural resources, such as air, water, and soil, all living organisms (animals, plants, people), and the built environment. The major ecosystems on the earth include soil, grasslands, savannas, shrublands, deserts, forests, tundra, freshwater, and marine.

Humans are one living organism that have interactions with all forms of ecosystems. Humans have inculcated the ability to alter organisms and interfere with the energy flows within an ecosystem. In this way an ecosystem is altered as a whole. Since the dawn of humanity, humans have been supported by their environment and at the same time altered it along the way. The earliest ancestors of homo sapiens lived approximately 4 million years ago and had the least destructive effect on the environment. They were hunters and gatherers and lived a subsistence lifestyle, utilizing what was available in nature for their survival. Overfishing, overhunting, and depletion of immediate resources prompted small nomadic groups to move to new areas where life could continue to be supported. The abandoned ecosystem had time to replenish.

When farming developed in the Tigris and Euphrates River Valley and in China around 8000 B.C., the humankind was at the early stage of initiating changes in the environment that would more greatly impact the ecosystem. To fulfil their needs, humans began to alter the environment by diverting water from rivers, adapting plants and animals through domestication, and settling in one place and building towns. As production of food increased, the size of the population also increased and specialized jobs developed. Now not everyone needed to be occupied with hunting and gathering, and growing food. This allowed urban and complex civilizations to develop that over time and with scientific discoveries blossomed into highly technological and industrialized world which we know today.

Never before has humanity used such ingenuity to utilize natural resources to create myriads of inventions, consumer products, and objects as they exist in our world today. In this process humans have depleted the natural resources utilized to make these things, disturbed ecosystems in pursuit of production of these goods, created pollution of water, soil, and air and produced greenhouse gases (GHGs) when manufacturing, transporting, and selling these items. A massive problem of waste disposal is also created when these items are not in demand anymore. Apart from this, modern advancements in technology has made it possible to support the ever-growing population. Of earth's nearly 8 billion inhabitants, India alone supports over 1 billion people.

What are the consequences of this modern way of life on the environment and the earth's ecosystems? Can nature continue to adapt to all these alterations induced by humans? Are there limits to how many people the environment can support? Natural resources such as water and trees are renewable, meaning they can be regenerated relatively quickly to meet human needs via normal ecological processes or agriculture. Others such as fossil fuels and minerals like gold and iron are non-renewable, that is, once they have been extracted from the earth they cannot be regenerated quickly. If even renewable resources are utilized or polluted at a rate that exceeds the carrying capacity of the environment, then they also cannot be regenerated. Evidence of this is quite readily available.

1.2 LOCAL TO GLOBAL SCOPE OF ENVIRONMENTAL ISSUES

Humans affect the environment at many levels. Climate change, acid rain, ozone layer depletion, overgrazing, overfishing, genetic engineering, species extinction, desertification, mining, and rising sea levels are just some of the effects of increased human impact on the environment. The ones most affected and of immediate concern are the homes and the local town or city. At this level, everyday actions and awareness can have immediate effects.

It is a well-known fact that if humans at individual level change day-to-day habits and actions that have adverse effect on the environment, changes can be witnessed at the global level. Steps taken at the individual level will include consumer products utilized, how one transports oneself, food eaten, energy and water consumed, and waste generated. At a city or town level, this would include how well people design and regulate infrastructure that is environmentally friendly.

On a national level, one must look at all of these issues as well as impacts of urban versus rural living, how much nature is being conserved and preserved on the whole, and active attempts at plant and animal species preservation. The complete impact can be felt in a nation as one begins to see overcrowding, urban sprawl, loss of habitat, endangering of species, polluted waterways, waste management issues, and public health problems.

On a global level, we need to examine similar and even larger cumulative effects of human action. When one considers the environment on a national and global level, however, the cumulative effects of every country's and the world's citizens gets harder to calculate. Climate change and its effects on weather patterns, glacial melting, and extinction of species are the best examples. These larger cyclical changes are harder to understand and predict although scientists have made many observations and are tracking many irreversible changes. More industrialized nations like the U.S., Western Europe, and China are seen as larger contributors to climate change, while the impacts tend to affect the poor more detrimentally, irrespective of wherever they live.

1.3 DEFINITION, SCOPE, AND IMPORTANCE OF ENVIRONMENTAL STUDIES

While it is challenging to understand, untangle, and alter the adverse impact of human action on environment, it is imperative that the awareness of these environmental concerns from the local to the global level is transmitted to people with a sense of urgency. The ethics of living sustainably, that is, living in such a way that humans do not compromise the natural environment to the point of destruction while also preserving human life, culture, and economic vitality must be nurtured across the globe and at all levels of society.

A multidisciplinary approach that combines all of the sciences and social sciences is needed in order to deeply understand the origins of environmental issues, their complexity, and their impact. *Environmental studies* seek to do just that. All fields of natural science, that is, biology, botany, chemistry, geology, hydrology, physics, and zoology can shed light on the workings of the natural world and the impacts of human activity on this natural environment. All forms of social sciences and historical analysis, that is, anthropology, archaeology, civics, economics, geography, political science, and psychology can be useful in understanding how human impacts on the natural environment have varied over time and among cultures. These disciplines can also inform us of ways to proactively affect change at the individual, local, national, and global levels. These multifaceted disciplines and understandings they convey relationally are known as *environmental studies*.

The Precautionary Principle highlights the thoughtfulness that an environmental studies course which integrates both the natural and social science seeks to promote. Many environmental advocates utilize the Precautionary Principle adopted by UNESCO (the United Nations Environmental, Social, and Cultural Organization) in 2005 as a basis for thinking about environmental issues at the global level. This principle, in recognition of current serious environmental threats, calls on society to exercise caution and prudent restraint in environmental decision-making and the production and consumption of resources. The principle also calls for decision-makers to meet all humans needs (not just for those in the most industrialized and wealthy nations and classes of society) and to protect the rights of workers, local communities, and the general public, in a manner that causes the least environmental harm. This principle is a noble challenge and call to action for all.

Thus, the scope of environmental studies is broad and purposeful. It seeks to help humans live more consciously and aware of the natural resources used in their life and the impact of their actions individually and collectively on the environment. With a firm knowledge base, such educated citizenry can confront the magnitude of current environmental issues caused by humanity. As these issues grow, change, and are debated in public, environmental studies can foster solutions that will ensure the future of all life on earth.

Environmental studies seek to understand the relationship between humans and the natural environment. The importance of such studies cannot be underestimated in our environmentally challenged world today. Global warming due to greenhouse effect and rise in temperatures around the world are forcing a global debate on how the resulting climate change will impact populations around the world. Unless we reduce the release of carbon dioxide and other GHGs into the atmosphere, climate changes will impact more vulnerable populations and low-lying coastal communities will be submerged. This book discusses the current status of climate change and the mitigation measures that can be undertaken to minimize its adverse impact.

This book summarizes the major topics in environmental and ecosystem sustainability. Projects that can be completed in college and in the field are detailed to give the student hands-on familiarity with environmental issues. The importance of environmental issues in society is described at length with several examples from India. Two chapters are dedicated to importance of environmental education and public participation. A unique feature of this book is the wealth of International Case Studies where lack of environmental considerations has caused major adverse impacts in society.

2

Sustainable Ecosystems

2.1 INDRODUCTION

An ecosystem is defined as an interaction between living beings and the non-living environment in which they exist. Ecosystems exist on many scales: a large ecosystem contains many smaller ones. Also, one ecosystem may look quite different from another. The community of animals, plants, insects, and other organisms that have adapted to a high mountain range is different from the community of living beings that have adapted to a tropical forest, ocean, or desert. However, all ecosystems have a food web, nutrient cycles, and energy flow, and each part of the system is connected to another. Every ecosystem has a measurable carrying capacity: over-expansion or die-off of one species or disturbance of an important natural resource can affect the sustainability of the whole system. Pollution of water, air, and land and habitat destruction of animals can significantly affect the carrying capacity of an ecosystem. Climatic variations due to greenhouse gas emissions are harming the ecosystem. Governments around the world have been trying to minimize these variations by reducing the usage of fossil fuels and replacing them with renewable energy sources such as solar, wind, and bioenergy.

2.2 UNDERSTANDING ECOSYSTEMS

Every constituent of an ecosystem, whether living or non-living, has an important role to perform (Figure 2.1). For instance, forested slopes prevent erosion, trees produce oxygen and increase moisture in both air and soil, birds carry seeds to new locations, insects live on tree branches and eat away fungus, and microorganisms on the forest floor recycle leaf litter and other organic matter into nutrients that help the trees to grow.

Ecosystems are of different types: terrestrial—deserts, forests, grassland, tundra, mountains, islands, savannas; aquatic—oceans, lakes, ponds, rivers; and ecosystem where land and water meet—shores and swamps. Farms and cities are examples of human influenced ecosystems. Further, garbage dumps, parks, and sewers in cities form ecosystems with their own communities.

Figure 2.1 An ecosystem connects all other creatures in the community

An ecosystem is made of both living (biotic) and non-living (abiotic) components. The interactions between biotic and abiotic components affect an ecosystem's equilibrium and survival. If one part of an ecosystem changes, it will affect all the other parts of the system. Ecosystems have feedback loops that make them more stable.

Bacteria, fungi, plants, insects, and animals, as well as waste products of these living things, such as fallen leaves and branches from plants and faeces, urine, and dead bodies of animals, form the biotic components of an ecosystem. The abiotic components include sunlight, water, nutrients such as carbon, hydrogen, oxygen, nitrogen, phosphorus, and sulphur, and elements of climate such as rainfall, temperature, moisture, yearly cycle of weather patterns, salinity of water, and wind patterns. These factors have a great impact on the evolution, behaviour, and populations of living things in the ecosystem. All these factors help determine how the parts of the ecosystem will interact. Each member of the ecosystem community, from fly to tiger, occupies an important niche that helps to organize the system.

Sun is the main source of energy in an ecosystem. Whether in tropical or cold climates, summer is the most productive period because it is the season that receives the most sunlight. Longer summers or cooler summers owing to man-made climate changes affect the energy in the ecosystem and result in undesirable effects such as extinction of some species.

Ecosystems change over time in a process known as ecological succession. In this process, the dominant species of the ecosystem changes over time in response to gradual changes in energy or resources, or in response to a sudden disturbance. For instance, a grassy meadow kept mowed by grass eaters may gradually turn into a forest if the grass eaters are hunted off it. A forest may have very little undergrowth because of a thick canopy blocking light at the ground level and the producers there not getting enough energy to reproduce in large numbers. When trees fall—due to age, hurricane, fire, or human logging—light reaches to the ground level and new grasses and shrubs may thrive.

2.3   ECOSYSTEM FUNCTIONS

Ecosystem functions, also known as ecological processes, are the biological, geochemical, and physical processes that take place or components that occur within an ecosystem. These exchanges or interactions between the parts of every ecosystem (example, vegetation, water, soil, atmosphere, and biota) ensure the survival of living organisms in the community and manage the flow of energy and nutrients that support life. For instance, plants capture energy from sun through the process of photosynthesis. Using sun's energy to grow, plants produce biomass, which animals and insects eat. In this way, energy of the sun is transferred from plants to insects and herbivores or plant-eating animals.

Animals higher up the food chain eat the herbivores and insects lower down. After the insects and animals die, microorganisms, including fungi and bacteria, decompose the leaf litter and animal bodies, breaking them back down into the basic nutrients that plants require to grow.

Ecosystem functions that specifically benefit humans are sometimes called "ecosystem services". For instance, natural systems filter pollution out of water. Bees pollinate crops, marshes protect shorelines from flooding, and green spaces produce oxygen and remove carbon dioxide from atmosphere. Apart from providing timber for our houses, cooking fires, and furniture, trees also provide habitat for birds and other animals, as well as make the air cleaner and healthier by absorbing carbon dioxide and emitting oxygen, which prevents global warming. In recent times, global warming has gained a lot of importance, and the Kyoto Protocol in 2007 and subsequent agreements have provided funds for reforestation.

Ecosystem services are crucial to human survival and are one of the reasons why it is important to preserve the natural environment. If humans had to artificially recreate all the services that the natural world provides, it would be enormously expensive, if not impossible.

2.4   PRODUCERS AND CONSUMERS

Within an ecosystem, every organism has a function to perform. Producers, consumers, and decomposers are the three main functions that organisms play in an ecosystem (Figure 2.2). Producers can perform photosynthesis or absorb energy from sunlight and transform it into carbon molecules in forms that other animals can use, such as carbohydrates. Primary producers make their own food from sunlight and are called autotrophs. The energy they produce from sunlight is the source of all energy in the ecosystem other than heat. All plants, as well as algae and photosynthesizing phytoplankton in the oceans, are primary producers.

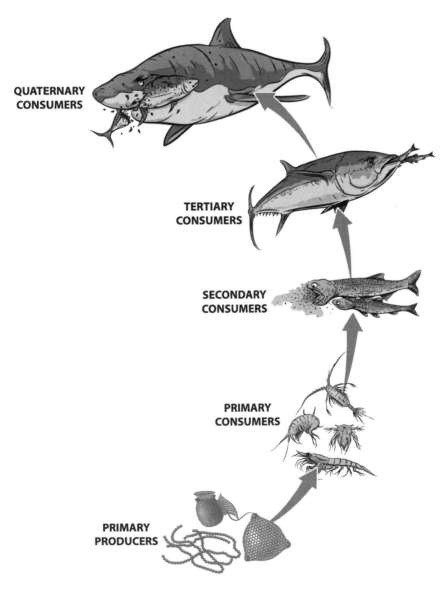

QUATERNARY
CONSUMERS

TERTIARY
CONSUMERS

SECONDARY
CONSUMERS

PRIMARY
CONSUMERS

PRIMARY
PRODUCERS

Figure 2.2 Producers and consumers in the ecosystem

Consumers are all those members of an ecosystem that cannot produce energy from the sun but must get their energy by consuming those that produce energy. Primary consumers are organisms that eat producers. They are also called herbivores or plant eaters. Any plant-eating animal is a primary consumer. Fish that eat algae are also primary consumers.

Secondary consumers are those organisms that eat primary consumers. They range from birds that eat plant-eating insects to people who eat herbivores such as goats or algae-eating fish.

Tertiary consumers are those animals that eat carnivores. These include carnivorous fish (fish that eat other fish) and animals. The different levels of consumers based on what they eat are called "trophic" levels.

Some toxic substances such as mercury and other heavy metals are difficult for animals to excrete. These toxins may exist in the environment in levels that are not very harmful but can get concentrated in animals at high trophic levels further up in the food chain to the point where they are poisonous, as they eat other animals that have the toxin in their bodies. This process is known as *bioaccumulation*. Mercury is a toxin that often builds up in the bodies of fish that people eat. Big carnivorous fish may contain so much mercury that people can get poisoned by consuming them frequently. In the 20th century, hawks and eagles became threatened with extinction worldwide after accumulating the pesticide DDT in high concentrations, which prevented them from forming hard eggs. The threat of bioaccumulation to high-level consumers, including humans, is one of the reasons why we need to regulate industrial pollutants.

Only about 10% of the energy in each trophic level is passed to the next level. That is why there are more grasses than there are deer or antelopes that eat the grasses, and there are more herbivores than carnivores such as lions or falcons at the top of the food chain. These carnivores need a much bigger range to hunt for their prey.

The loss of energy between trophic levels is also the reason why it is more energy efficient and environmentally friendly to be a vegetarian than a meat eater.

Decomposers, or detritivores, break down organic molecules that are no longer living, such as dead leaves, plants, and animals, as well as excretions of living organisms, to their abiotic pieces. Decomposers such as bacteria and fungi eat only dead organisms. Mushrooms on a log or mould on a piece of rotting fruit are examples of decomposers at work. Decomposers perform the important task of breaking down complex organisms into carbon, nitrogen, phosphorous, and other basic abiotic ingredients so that they can be recycled into new growth in the future.

2.5 ECOSYSTEM PROCESSES

The most important processes in an ecosystem are the energy cycle, the water cycle, and the nutrient cycle. The connections that define an ecosystem are meant to ensure that many of the system's building blocks are passed from one part of the system to another to keep the whole system functioning. The water cycle is shown in Figure 2.3.

2.5.1 Energy Cycle

The sun is the ultimate source of all energy that enters an ecosystem. It is an abiotic element of the ecosystem. Sun's energy enters the system as light and is captured and transformed to chemical energy through the process of photosynthesis. Plants use solar energy to convert the nutrients in soil to sugars, carbohydrates, proteins, fats, and other organic molecules. This food is used by animals and humans to derive energy.

According to the first law of thermodynamics, energy cannot be created or destroyed. It can only be changed from one form to another. Some energy is lost as heat when it is transferred from one organism to another.

Even though energy cannot be created or destroyed, it can enter or exit an ecosystem. Energy that enters the system as light can escape the system as heat. Some energy is lost as heat when an animal breathes,

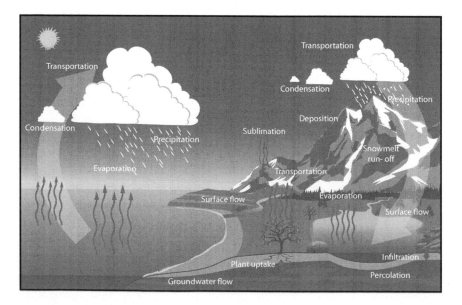

Figure 2.3 Three phases of water cycle—evaporation, precipitation, and transportation—keep water flowing through the living world

moves, or performs other functions necessary for living. Energy is an "open system" because it can enter from the sun or be lost as heat. Energy flows through the system as one organism consumes another. The predator–prey relationships in the ecosystem are the main driver of energy flow, with the successful predator receiving energy from the food provided by the prey.

In addition to flowing from one creature to another during consumption, organic molecules created from sun's energy may also flow through the system by being recycled into their inorganic components through decomposition. Microorganisms such as bacteria and fungi break down dead plants and animals or excretions of living animals to abiotic elements and return the carbon and nutrients back to the ecosystem.

2.5.2 Water Cycle

Water moves through an ecosystem, from clouds to rains to rivers, in a process called the water cycle. It flows through the system via evaporation, condensation, and transpiration.

Evaporation occurs when water changes from a liquid to a gas or vapour and rises from the earth's surface to the atmosphere. Water evaporates more quickly in warm temperatures; warmer air holds more water than cooler air. Evaporation takes up energy, cooling the surrounding environment. About 90% of the water in the atmosphere comes from evaporation.

Condensation brings water back from the air in the form of clouds, dew, rain, and snow. It is affected by differences in temperature, just as water in the air condenses on a cold surface. While evaporation cools the surroundings, condensation warms the surroundings.

The remaining 10% of water in the air comes from plant leaves through the process of transpiration. The amount of water produced by plants through transpiration varies depending on the temperature, humidity, wind, and moisture available in the soil.

Water affects the climate in large and small ways. Taking up and releasing heat during evaporation and condensation, respectively, water helps distribute heat around the globe. Because white colour of ice reflects sunlight, it helps send some of the sun's heat back into space. As the earth's warming climate melts the Arctic ice, the dark water that replaces the ice absorbs heat instead of reflecting it, thus magnifying the effect of global warming. A warmer climate speeds up the water cycle, causing more evaporation and heavier rainfall. It can also shift rainfall patterns, creating problems for farmers.

After the rains, water moves through the ground as run-off, carrying minerals, watering natural ecosystems and agriculture, and filtering pollutants. Forests, marshes, and other greenery minimize run-off, helping ecosystems retain moisture. Dry lands can sometimes turn into deserts, a process known as desertification, when livestock overgraze and trees are cut for firewood. Improved land management can help build healthy ecosystems that make the most of the water cycle to ensure that land can support life for both humans and other creatures on earth.

2.5.3   Nutrient Cycle

The nutrient cycle describes how ecosystems move important abiotic elements, such as carbon, nitrogen, phosphorous, sulphur, and potassium, from the physical environment to living beings and back to minerals that other living beings can use.

The physical environment contains nutrients that are unevenly distributed because of the geology or use of land. Ecosystems help move nutrients to living beings. They cycle nutrients through the food web. Plants use nutrients to build biomass, which supports varied life. Decomposers then recycle these materials, breaking them back so that they can be reused.

Some plants are good at nutrient cycling. Leguminous plants such as lentils increase the nitrogen content of soil. Farmers spread manure on their fields and in the process recycle the carbon and other organic matter back to the soil so that crops grow better.

Artificial movement of nutrients, such as chemical fertilizers, has tremendously increased crop yields, but this can also disrupt ecosystems. If too much nitrogen enters lakes and oceans from farm fields, it can cause loss of life in huge areas of the waterways. Humans can help recycle nutrients to renewable resources, for instance, by composting food waste to fertilizer rather than disposing of in a toxic landfill. Managing nutrient cycles sensitively can help make ecosystems on earth more sustainable.

2.6   FOOD WEB

The food web describes how the living members of an ecosystem feed themselves, from the smallest microorganisms to the largest mammal (Figure 2.4). As discussed previously, energy is transferred throughout an ecosystem when one organism feeds on another. From primary producers to apex or tertiary consumers, each creature occupies a different trophic level.

A hierarchical view of this energy transfer is called food chain: big fish eat small fish, which eat tiny fish, which eat plankton and algae.

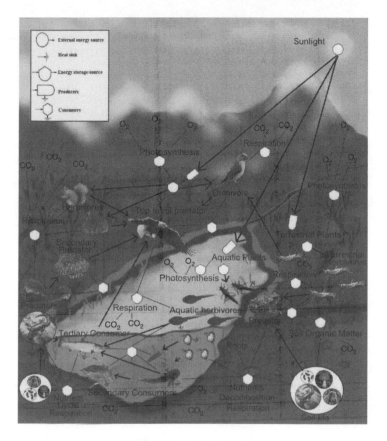

Figure 2.4 Food web

Energy is lost at each level of transfer, which is why swordfish are fewer in number than minnows. On land, there are more grasses than herbivores and more herbivores than carnivores.

A more multidimensional view of food in an ecosystem is a food web. In this view, we find that not only bigger fish eat smaller fish, but birds, bears, and humans also eat fish. Many animals, including snakes, monkeys, and humans, eat birds or bird eggs.

There are more individuals at the bottom of the food chain than at the top, and their survival strategies are based more on numbers than on preservation of individuals. Throughout the web, populations are affected by the available food supply. If grasses have a good year, then the population of grass eaters may explode. The carnivores that feed on grass eaters may then have healthy babies with a better chance of survival.

The food web also helps ecosystems regulate populations through checks and balances. If plentiful food for prey explodes the population of predators, the prey population may be thinned out, followed by the decline of the predator population because of the trouble in finding prey. If predators—even apex predators at the top of the food chain, such as tigers—are killed by humans, the balance of the ecosystem may be disturbed, allowing too many prey species to survive.

For example, people in the United States hunted coyotes, foxes, and wolves to extinction in many places. As a result, the deer and rabbit populations then exploded, and the entire forest ecosystem was affected by the eating habits of these herbivores.

Food webs in ecosystems thrive on biodiversity. These webs may be disrupted by habitat loss, such as when humans replace diverse natural ecosystems that sustain many species with agricultural fields growing only a single plant species. But ecosystems are healthier when many species fill the same niche. The system can then better adapt to disease, population fluctuations of predators or prey, or weather that varies from one year to another. Nature's food webs are a way of verifying how all life forms on earth are connected to one another.

2.7   INFLUENCE OF HUMANS ON ECOSYSTEMS

Humans influence ecosystems in many ways (Figure 2.5). For instance, humans change habitats: animals that thrive in a meadow may die if the meadow is turned into a wheat field. Plants and animals that evolved for millions of years in large forests may become endangered when humans build houses, farms, and cities where the forests used to stand.

When there are disruptions in an ecosystem, such as depletion of resources, natural disasters, over- or underpopulation of a species, or other stressors such as climate change, organisms in the ecosystem must adapt and change to survive. While humans are only one of millions

Figure 2.5   Humans have a huge impact on an ecosystem's ability to remain sustainable

of living species that populate the planet, they have a large impact on ecosystems around the globe.

The human population on earth has grown from less than 400 million only a thousand years ago to more than 7 billion today. India is one of the most densely populated regions on earth. While India covers only 2.4% of earth, it contains more than 17% of the world's people. When fewer humans lived on the planet, human consumption and their impact on global ecosystems were limited. However, with rapid growth in population, humans are using the earth's resources in ways that adversely impact the ability of ecosystems to support life.

For instance, ecosystems evolved over millions of years to have checks and balances that make them fairly stable. If insects eat too many plants, soon there will not be enough plants to feed for insects and the insect population will drop. If there is not enough water, then fewer plants will grow and the demand for water will decrease. Humans can eat a variety of foods, live in nearly any kind of climate and habitat, and change ecosystems to suit their needs. For instance, humans have converted so many forested areas to residential places, farms, and factories such that at many places not enough pristine forests remain to support the ecosystems that evolved there.

Since many species on earth are adapted to live only in limited ecosystems—on a high mountain or in a clean river delta—the existence of many species and entire ecosystems may be threatened by human activities, ranging from mining to water pollution from factories. Conservation experts estimate that as much as a third of species on earth are threatened by human development.

Humans have been influencing even the biggest ecosystem of all, the global climate system. We are burning so much fossil fuels that carbon levels have steadily risen in the atmosphere, which is raising global temperatures and causing alarming changes in weather patterns, including monsoons and drought. The human impact on cycles of nutrients and natural food webs has become so drastic that it threatens human survival as well.

Poverty, famine, water shortages, wars over natural resources such as petroleum and water, and many health problems are made worse by the pressure humans are putting on their ecosystems. While natural systems flow in cycles, human-engineered systems often go in one direction: extraction, consumption, and disposal. The combined health of all ecosystems on the earth will ultimately determine human habitability, as humans are but one part of the earth's extensive webs and cycles.

Wasteful consumption and overconsumption have many negative effects on our ecosystems, such as the following:

- Loss of forest lands, lack of reforestation, loss of biodiversity
- Degradation of waterbodies by municipal and industrial pollution
- Filling of wetlands for building houses and industrial facilities disrupts the nature's ability to manage the water cycle and causes flooding.
- Overfishing of oceans and lakes disrupts the delicate balance of marine life.
- Overgrazing and intensive farming lead to desertification of productive land.

Natural ecosystems can remain in balance if humans conserve their natural resources, reduce wasteful consumption, and manage resources in a sustainable manner.

2.8   TYPES OF ECOSYSTEMS

Ecosystems can be classified into many types based on the climate, soil type, altitude, and many other factors, but four major systems described in this chapter are forests, grasslands, deserts, and aquatic ecosystems. Each of these systems is unique and provides resources for human survival. These major ecosystem types are distributed in many parts of the world. Figure 2.6 shows the location of these ecosystems in various parts of the world.

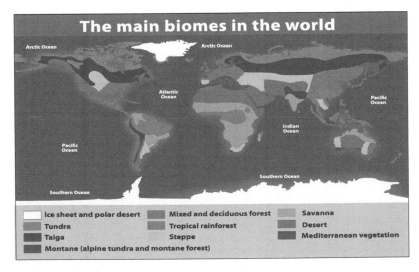

Figure 2.6   World's ecosystems can be grouped into "biomes", which share many characteristics whether they exist in Canada or Siberia, Africa or India

Over long periods of time, ecosystems may go through stages in a process known as "succession", in which a grassland slowly evolves into a forest, or dessert ecosystem, depending on the climatic changes.

2.8.1   Forest Ecosystems

Trees are the dominant community in a forest ecosystem, and they play a vital role in controlling the energy, water, and nutrient cycles in the ecosystem. In an undisturbed, old-growth forest, the trees may be huge with relatively little undergrowth because the thick forest canopy blocks light.

India is situated on tropical latitudes and evolved with several types of forest ecosystems. Forests in India have been much altered due to human activities in the last few thousand years. The forests found in India can be grouped into five major categories: moist tropical, dry tropical, montane subtropical, temperate, and alpine.

Forest plants vary considerably depending on climate and soil conditions. India's moist tropical forests are populated by wet evergreen, semi-evergreen, and moist deciduous plants, as well as swamp communities. The dry tropical forests are inhabited by deciduous, thorn, and dry evergreen trees. The montane subtropical forests have broad-leaved trees, pine, and dry evergreen trees. The montane temperate forests contain wet, moist, and dry trees. The alpine or mountain forests of the Himalayas consist of sub-alpine trees on the lower elevations, moist alpine forests, and dry alpine trees, until the tree line at a high latitude where tundra takes over and trees can no longer survive.

2.8.2   Grassland Ecosystems

In some parts of the world, such as the Savannas in Africa or the steppes in Mongolia, grasslands have always existed, but the grasslands in India are not of primary origin. Grasslands developed in many areas of India due to the destruction of natural forests, particularly due to land clearing for agriculture and grazing. According to the moisture present in the area, Indian grasslands may be categorized into three types:

- Xerophilous (adapted to dry, hot conditions) grasslands, found in semi-desert areas of north and west India.
- Mesophilous (growing best at moderate temperatures) grasslands or dry savannas, found in Uttar Pradesh, having moist climate.
- Hygrophilous (preferring very moist ground) grasslands or wet savannas, found in regions of India with higher rainfall levels.

2.8.3 Desert Ecosystems

Deserts constitute one-fifth of the earth's land surface. These are defined by their extreme temperatures and very low rainfall levels. Animals that adapt to desert ecosystems learn to manage water carefully or store it over long periods. For example, cactus and other succulents store a lot of water so that they can handle infrequent rainfall. Camels also store water in their humps so that they can travel long distances without drinking.

Reptiles are cold blooded and so benefit from the warmth of the desert sun. Deserts seem relatively empty of both plants and animals because the number of creatures living in desert conditions is limited by the harsh conditions. But even among deserts, conditions vary. The Sonoran Desert of North America has quite a lot of animal and plant life, for instance, compared to the naked dunes of the Sahara.

In some deserts, a rich ecosystem evolves around a rare source of water, called an oasis. Examples of deserts around the world are the Sahara in Africa, the Sandy Desert in Australia, and semi-arid deserts such as the Sagebrush desert in North America.

Cold areas with little rainfall can also be considered deserts. Cold deserts exist in Antarctica and Greenland. Deserts cover about 17% of the earth's landmass, and their relatively low productivity is due to low rainfall resulting in low abundance of wildlife. Even though numbers are low, considering the harsh environment, biodiversity is relatively high. One-quarter of all global vertebrate species are found in deserts.

2.8.4 Aquatic Ecosystems

Aquatic ecosystems include different kinds of water environments. Lakes, rivers, floodplain marshes, and oceans are all types of aquatic or water-based ecosystems. Aquatic ecosystems contain a wide variety of life forms, including bacteria, fungi, and tiny organisms called plankton, as well as fish and marine mammals. Ecosystems at the bottom of deep ocean trenches are quite different from those in ponds and streams, but both are considered aquatic. Aquatic ecosystems can be classified into three major categories: freshwater, transitional communities, and marine or saltwater.

Freshwater ecosystems cover only 0.8% of the earth's area. One major type of freshwater ecosystem is standing water communities, known as "lentic" systems. Lentic systems can be of any size, ranging from temporary puddles to ponds and lakes. The other main type of freshwater system is "lotic" or moving water communities. Lotic systems include small streams and all kinds of rivers. While lentic systems tend to be quite local and contained, lotic systems can run for thousands of miles through many types of conditions. The life forms adapted to these two types of freshwater systems are, therefore, quite different.

The organisms that live in a standing water or lentic community includes algae, rooted and floating-leaved plants, invertebrates such as worms and filter feeders, crustaceans such as crabs, shrimp, and crayfish, snails and molluscs, amphibians such as frogs and salamanders, reptiles such as turtles and water snakes, and many species of fish.

The life forms in moving water or lotic ecosystems vary a lot depending on the strength of water flow and the amount of light penetrating the water. Deep and quiet areas in slow-moving rivers can be similar to still water (lentic) systems. The primary producers in lentic communities are algae. These provide food for other lotic community members such as fishes, crustaceans such as crayfish and crabs, aquatic insects, and molluscs such as clams and limpets. Bacteria in the water decompose dead leaves and other organic materials back to their components to be recycled in the circle of life.

Transitional communities are halfway between land and water. These include estuaries, riverbanks, marshes, and wetlands. Estuaries are the reaches of rivers where the freshwater from rivers reaches the ocean, blending with the seawater. The organisms living in estuaries can tolerate varying levels of salt in the water; the salt level may change with the tides. Estuaries are protected from the full force of ocean waves and from the fastest rushing parts of rivers, and so they often contain a rich diversity of aquatic life. Transitional communities are often threatened by human development.

The various types of transitional communities include wetlands, bogs, fens, swamps, and marshes. They perform important functions for both nature and humans, including filtering pollutants from water, serving as nurseries for fish, and buffering waves to help protect shorelines

Figure 2.7 Coral reefs are biodiversity hotspots

from flooding. Wetlands contain a rich diversity of species due to the abundant sunlight available to them. Wetland plants include water lilies, mangroves, tamarack, and sedges. Many species of reptiles, amphibians, and shorebirds also inhabit wetlands.

Oceans cover an estimated 71% of the earth's surface. Marine ecosystems encompass shorelines, estuaries, bays, coral reefs, and the open ocean, from the light-filled upper waters to the cold, deep bottom of the sea. Relatively shallow saltwater near the shore provides habitat to many species of burrowing invertebrates, such as clams, crabs, and worms. Life forms in the ocean are varied from the small plankton and krill to marine mammals such as gigantic whales and dolphins. Deep ocean water has high pressure and only a little light energy, and so the few creatures that exist there are specially adapted to these conditions. At the bottom of the ocean, some vents bring hot gases into the ocean from deep inside the earth. Ecosystems that evolved around these deep-sea vents take energy from the heat as there is no light.

Coral reefs are formed by the accumulation of calcium carbonate deposited by marine organisms such corals and shellfish. The clear, warm water in coral reefs supports a huge number of fish and other organisms. Coral reefs exist in only 1% of the oceans, but 34% of ocean species live in coral reefs (Figure 2.7). That is why reefs are considered one of the world's biodiversity hotspots, places where a large number of species are concentrated in a small area. Unfortunately, corals can live only in a narrow temperature range, and the warming of oceans in the age of climate change is threatening their survival.

Open oceans are an important heat sink for the earth's climate. They have a great impact on the biosphere and are the source of much of the rain that falls over land. Ocean temperatures have a big effect on climate and wind patterns. Oceans absorb a large percentage of the heat added to our atmosphere by global warming.

2.9 CONCLUSION

Sustainable ecosystems help the living world to function in a predictable, organized way. Ecosystem communities contribute to the stability of our world by cycling nutrients and energy and by forming the stable landscapes with which humans and other living creatures have evolved, knowing what to expect. Human disturbance, whether through diverting too much water, disrupting food chains with toxic pollution, or clear-cut logging, can upset the balance of nature. Sustainable ecosystems need to be respected so that they can continue to provide the valuable services on which our lives depend.

3

Natural Resources

3.1 INTRODUCTION

Planet earth is endowed with rich natural resources, such as water, forests, minerals, and productive land. While water is essential for life, forests and vegetation are essential for absorbing carbon dioxide emissions and releasing oxygen into the atmosphere. They also serve as sink for storm water and minimize flooding in urban areas.

The use of more natural resources is inevitable for supporting the growing population around the world. However, the effective and sustainable use of these resources would leave sufficient resources available for future generations. Equitable use of natural resources can improve the standards of living of global citizens and also sustain the earth's ability to support its inhabitants in the future.

Unfortunately, humans are destroying forests to acquire more agricultural land. The process of urbanization has also depleted forests and is contributing to climate change as well. Mining mineral resources without adequately protecting the environment impacts the communities living around mines. So how should India manage and preserve its natural resources for future generations?

This chapter discusses the laws India has passed for preventing the depletion of natural resources. It explains how to manage water for public health and safety, agriculture, industry, and ecosystem, as well as the costs involved. It describes various integrated water management systems that treat water supply and wastewater management comprehensively. The chapter also discusses sustainable management of forest and mineral resources.

3.2 WATER RESOURCES

Water covers about 70% of the earth's surface, but all the earth's water is not suitable for human needs. Only about 3% of the total water is available for drinking and other needs. Water is needed for drinking, industrial use, and agriculture. It is also vital for maintaining ecosystems that are essential for our well-being. Unless there is an understanding of how water and natural resources affect humans and ecosystems, these resources will not be managed sustainably.

3.2.1 Hydrologic, Nutrient, and Carbon Cycles in Water Use

The hydrologic cycle starts with rain, which provides the freshwater found in rivers, lakes, and underground aquifers. This water reaches the oceans through rivers. The sun evaporates the water from waterbodies and oceans to form storm clouds that lead to rain, thus completing the hydrologic cycle (Figure 3.1). Hydrologic cycle produces extremes of water: floods due to excessive rain and drought when rain stops for many months. These extremes have to be managed to prevent the loss of human and animal life as well as property.

Wastewater generated by animals and humans contains nutrients such as potassium, nitrogen, and phosphorus. These nutrients come from the food consumed by animals and humans, and the food comes from the cultivated land. The nutrient cycle can be completed by returning the nutrients in wastewater to land. By completing the cycle, improved plant growth and agricultural productivity can be achieved. Section 3.2.3 of this chapter discusses the methods of recycling the nutrients in wastewater and completing the nutrient cycle.

The increase in carbon dioxide, methane, and other greenhouse gases (GHGs) in the atmosphere leads to global warming. GHGs can be reduced

Figure 3.1 Water cycle

in the atmosphere by completing the carbon cycle and sequestering the carbon in plants grown with recycled nutrients. Plants take in carbon dioxide, and by recycling wastewater for agricultural uses, the carbon cycle can be enhanced.

3.2.2 Water Use for Public Health

An essential prerequisite for ensuring public health is providing clean drinking water to urban and rural populations. The Government of India has enacted several laws based on the standards established by the US Environmental Protection Agency (USEPA) and the World Health Organization (WHO). These laws are enforced by the Ministry of Environment, Forests and Climate Change (MoEFCC) and the central and state pollution control boards.

In the United States, the US Congress passed the Safe Drinking Water Act (SDWA) in 1974 to protect public health by regulating the public drinking water supply.[1] Amended in 1986 and 1996, the law requires many actions to protect drinking water quality in various sources such as rivers, lakes, reservoirs, springs, and groundwater wells. The law authorizes the USEPA to protect tap water and requires all owners or operators of public water systems to comply with primary (health-related) drinking water standards.

The SDWA amendments in 1996 require that the USEPA consider detailed risk and cost assessments and the best available, peer reviewed science in developing the primary drinking water standards. State governments implement the regulations under the USEPA authorization and also encourage the attainment of secondary (nuisance-related) standards. The primary and secondary drinking water standards are discussed in the following sections. The USEPA also establishes minimum standards for state programmes to protect underground sources of drinking water from surface pollution or underground injection of polluting fluids.

The WHO has established international drinking water quality standards.[2] These standards protect the public health of populations around the world, but enforcement of these standards varies from country to country. The WHO drinking water quality standards are discussed in the following sections. Almost 3 billion people around the world do not have access to water that meets these standards and suffer from diarrhoea and other water-borne diseases. The UN Millennium Development Goals

---

[1] Safe Drinking Water Act (1974, 1996). Details available at <www.epa.gov/lawsregs/laws/sdwa.html>

[2] World Health Organization. Details available at <www.who.int/topics/drinking_water/en/>

(MDGs) suggested that all nations in the world should meet the WHO standards by 2015. The United Nations is conducting several programmes to build capacity in developing countries to achieve the MDGs.

3.2.3 Water Use for Public Safety and Recreation

The water required for fire protection should be available at all times. The USEPA has established guidelines on the quantity and quality of water for such public safety needs. Another major use of water is for recreation in our waterbodies, which should be fishable and swimmable. The Federal Water Pollution Control Act was passed in 1948 by the US Government to address the problem of pollution of waterbodies. The growing public awareness and concern for controlling water pollution led to a major amendment of this law, and the Clean Water Act (CWA) was enacted in 1972.[3]

The CWA establishes the basic structure for regulating the discharges of pollutants in the US waters and the quality standards of surface water. The CWA amendments in 1977 established water quality standards for surface waters and attained them by controlling the end-of-the-pipe discharges (point sources) from industries and municipalities. The point source discharges are regulated through the National Pollution Discharge Elimination System permits. Subsequently, the USEPA issued guidelines for non-point sources of pollution, which include discharges from farms, construction sites, and other land sources. By enforcing the regulations for point and non-point sources, rivers and waterbodies in the United States have been cleaned up over the years and most of them are fishable and swimmable, which promote public safety and recreation.

Although India has passed several water pollution control laws, the enforcement has been lax. The Yamuna River in Delhi is an example of how the government inaction has hastened the deterioration of the river's water quality. Only a small portion of the municipal wastewater is treated before being discharged in the Yamuna River. Industrial pollution from small- and medium-scale industries has rendered many sections of the Yamuna water unsafe for drinking, fishing, and swimming. Water quality measurements have shown that the Yamuna River is pristine until it reaches Haryana and Delhi, indicating clearly the impact of pollution on its deterioration.[4] Many other waterbodies in India have been severely affected by pollution. Unless the people demand that their water quality be protected by the government through the cleaning up of municipal

---

[3] Clean Water Act (1972, 1977). Details available at <www.epa.gov/lawsregs/laws/cwa.html>

[4] Rivers of the World Foundation. Details available at <www.rowfoundation.org>

pollution and enforcement of laws, poor people, who depend on this water for their drinking water needs, will continue to suffer.

3.2.4 Water Use for Agriculture and Industry

India is an agricultural country with about 80% of the available water being used for irrigation. Major conflicts have occurred between states over river water allocation for irrigation, and the Supreme Court had to intervene in such occasions. As Indian industry, including the power sector, expands to meet the needs of the population and the growing economy, there will be more demand of water. Given the water demand for agriculture and industry, municipalities do not have enough water to serve domestic needs, and severe water shortages occur in metropolitan areas. In Chennai, a power plant had to recycle municipal wastewater by treating it to the water quality needs and use it for generating 200 MW of power. The only way to balance the needs of agriculture, industry, and households is to recycle all the wastewater from municipalities and industries. In addition, recycling municipal wastewater for agriculture also recycles the nutrients and improves agricultural productivity. Rainwater harvesting coupled with wastewater recycling and reuse can meet all the water needs of the society.

Many communities in the United States initiated wastewater treatment and recycling after the USEPA launched a federal programme in the early 1970s to assist states with funding for treatment plants and for measures to prevent pollution of drinking water sources. For example, the Deep Tunnel Project in Chicago, Illinois, where a system of tunnels (109 miles) and two large reservoirs, helps in preventing wastewater from entering Lake Michigan (the drinking water source) during heavy storms.[5] Old cities such as Chicago have combined sewers that handle both rainwater and sewage. These sewers, along with the existing wastewater treatment plants, do not have the capacity to treat all the wastewater during the rainy season and the wastewater enters waterbodies. Building and maintaining costly water infrastructure are required to ensure that public health is not adversely affected by the pollution of drinking water sources.

India has more than 630,000 villages with a population of almost 700 million. Many of these villages do not have adequate drinking water supply systems and no wastewater treatment systems. Expensive irrigation systems have been constructed in many parts of India, and when a village lacks this infrastructure, it is totally dependent on monsoon rains.

---

[5] Metropolitan Water Reclamation District of Greater Chicago. 2008. Tunnel and Reservoir Plan for Chicago Metropolitan Area, Chicago, IL

Inexpensive wastewater treatment lagoons powered by solar or wind power for aeration can be installed in villages to recycle wastewater for agriculture and improve agricultural productivity, while enhancing the quality of the drinking water. Common effluent treatment plants (CETPs) have been installed in many industrial estates all over India to recycle wastewater and conserve freshwater. However, the operation and maintenance of CETPs are critical to ensure that polluted water from industries does not enter local waterbodies or groundwater resources. Many times CETPs do not work efficiently. Hence, monitoring and enforcement of regulations at these CETPs are essential to protect the water quality in the vicinity of industrial estates.

3.2.5 Integrated Water Management

A municipal agency plans for adequate water supplies and focuses on sources of water and how to distribute the water to residential, industrial, and agricultural users. If the plans are not coordinated with the wastewater agency, drinking water supplies get contaminated. For example, a septic tank located uphill of a drinking water well contaminates the well with the flow from the tank. Integrated water management has been advocated by the WHO. It involves planning the water supply and wastewater infrastructure in a manner that will minimize the overall costs, maintain the water quality, and allow for efficient recycling of wastewater. In urban and rural communities, integrated water management ensures that the treated wastewater is recycled efficiently through the judicious placement of water and wastewater treatment plants. In Los Angeles, there is very little rainfall and most of the water comes from the adjoining state through a costly infrastructure of pipes and pumping stations. The city decided to place the water treatment plant next to the wastewater treatment plant so that the treated wastewater flows by gravity to the water treatment plant. The water treatment plant filters the treated wastewater, adds a disinfecting chemical, and pumps the water to the city distribution system. This arrangement saves a lot of energy, and all water is recycled efficiently.

In Indian cities and villages, where water comes from surface water sources and adequate wastewater treatment facilities are not available, integrated water management can prevent wastewater drains from contaminating drinking water sources. Wastewater from sinks and showers can be recycled for gardening, and sewage can be treated using primary and secondary means. Drinking water sources (both surface and underground) should be protected from contamination and used only for drinking water supplies. Integration of rainwater harvesting and wastewater management will enable the most efficient use of water resources and prevent water shortages and flooding during storm events.

3.2.6 Costs of Water Infrastructure

Although water is a natural resource, its management for preventing contamination and providing clean water is a costly affair requiring decision-making and intervention. Water infrastructure can be divided into the following categories:

- Rain or storm water management
- Water treatment and distribution system
- Wastewater treatment and recycling system

3.2.6.1 Rain or storm water management

Rainwater can be harvested for immediate use or for replenishing groundwater sources. It can be harvested from our rooftops and stored in underground tanks or reservoirs (Figure 3.2). In Tamil Nadu, it is mandatory for every building to harvest rainwater for immediate consumption or for storing it in underground aquifers. This law has enabled the water table in Chennai to rise by 5 m and has solved the problem of water shortages in summer. Owing to the lack of proper management of storm water, flooding occurs in cities such as Mumbai and Delhi immediately after a heavy storm event. In addition to rainwater harvesting, extensive drainage structures are required to convey the run-off from streets and other paved areas to reservoirs or open areas where the water can percolate into the ground.

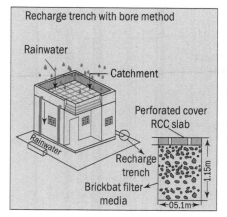
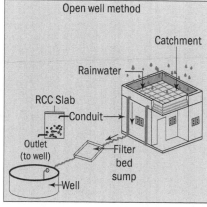

Figure 3.2 Rainwater harvesting

In the United States, drainage structures are planned and built along with the road infrastructure to prevent flooding during storm events. Such planning is lacking in India as proved by the recent disastrous floods in Mumbai and Chennai. Natural drainage features such as tree-filled detention basins, planted sidewalks and medians, and open areas are effective in managing storm water. Structural drainage facilities consist of pipes, collection basins, and earthen levees/embankments. Storm water collects the dirt, oil, and other debris on streets and, consequently, has to be treated before it is released to waterbodies. The best way to manage storm water in urban areas is to have low-intensity development and minimize paved areas. In older cities around the world, storm water and wastewater are handled through the same drains, which creates management problems.

3.2.6.2 *Water treatment and distribution system*

Water treatment removes particulates using sand and carbon filters and disinfects water to remove traces of biological contaminants. However, the cost of treatment and distribution to users is very high. Water quality is the most critical aspect in delivering clean water to the public. With improved knowledge about the impact of water on human health and measurement technologies, humans are capable of achieving better water quality. However, this has major cost consequences.

The permissible level of arsenic in drinking water was previously set at 50 parts per million (ppm), but now it has been reduced to 10 ppm. The importance of fluoride in drinking water was recognized a few years ago, and many municipalities in the United States add fluoride to drinking water to improve dental health. However, in many parts of India, excessive fluoride in drinking water sources is causing significant health problems. Hence, monitoring water quality and making costly improvements in water treatment plants are ongoing processes.

Distribution systems are expensive and have to be maintained to prevent leakage and loss of clean water. In the state of Illinois in the United States, there is a large body of drinking water in Lake Michigan. However, making this water available to the population living about 50 miles from the lake through large pumping stations and pipelines is very costly. In the late 1990s, the local and state governments spent more than $400 million to supply water from Lake Michigan to about a million consumers in the suburbs of Chicago.

Israel has a preventive maintenance plan to monitor and replace leaking water and wastewater pipes throughout the country, and this saves millions of gallons of water every year. However, such programmes are expensive and user fees are required to recover the costs of delivering water. Unless everyone pays his or her fair share for the delivery of water, water infrastructure systems cannot be improved or maintained. It is likely that in most nations, including India, the lower income sections of the society would have to be subsidized by the government since water is essential for the protection of the right to life.

The productivity of populations depends on the availability of clean water. Governments, either by themselves or through public–private partnerships, have to build and maintain the infrastructure for delivering clean water. A balance has to be maintained between surface and groundwater sources for drinking water since the depletion of groundwater aquifers would have long-term adverse consequences. Also, protection of these water sources from contamination is also costly and needs constant vigilance. Awareness about these issues is seriously lacking among Indians, and a concerted effort is required by public health agencies to improve the situation and deliver clean water to the public.

3.2.6.3 *Wastewater treatment and recycling system*

Wastewater is produced by industries and municipalities. Wastewater from industries has many harmful contaminants, such as radioactive metals and harmful chemicals. In municipalities, wastewater is mainly generated by humans and it can be treated and recycled with simple biological processes. Wastewater produced in industrial estates in India is treated at CETPs and recycled within the estate or discharged to waterbodies. However, CETPs are not present in all industrial estates and are inefficiently run in many places, thus creating significant pollution of waterbodies. In India, only a small amount of municipal waste is collected and treated before recycling and a majority of the municipal waste is directly discharged into waterbodies and rivers.

In the United States, wastewater is treated in publicly owned treatment works (POTW), which treat both municipal and pretreated industrial wastewater. In the last 35 years, thousands of POTWs have been built in the country and have significantly improved the water quality in rivers and waterbodies. The costs for such treatment plants range from a few million dollars for a small municipality to a few billion dollars for a large city such as Chicago. The key components of a municipal POTW are as follows:

- **Collection system:** It consists of small- and large-diameter sewer pipes to collect wastewater from homes and deliver it to the treatment plant. In addition, several pumping stations are required to deliver wastewater. In large cities that have combined sewers (storm water and wastewater), large tunnels are required to deliver the water.

- **Primary treatment system:** It removes solids and large particles from wastewater. This system reduces the biological oxygen demand (BOD) and is a minimum requirement in treating wastewater. The higher the BOD in wastewater, the lesser the oxygen available for fishes, which leads to fish kills. As per the MoEFCC regulations in India, the BOD of water discharged to waterbodies should be less than 30 mg/l.

- **Secondary treatment system:** It removes pathological contaminants such as faecal coliform bacteria from wastewater and makes it suitable for recycling. Biological treatment through aeration is the most common secondary treatment. It produces a lot of sludge, which is expensive to dewater for land application. However, simple lagoon systems for small, decentralized wastewater treatment systems can be designed to produce a minimal amount of sludge.

- **Tertiary treatment system:** It removes particulates through a filtration system and makes the wastewater suitable for recycling for drinking or industrial uses. Filtration systems can vary from simple sand and charcoal filters to the reverse osmosis method, which involves high-pressure filtration through membranes. This is a very expensive part of the treatment process, depending on the quality of water required after treatment. For power plants and other industrial applications such as pharmaceutical manufacturing, this treatment step is costly.

3.2.7 User Fees and Government Subsidies

The government or the private sector has to recover the huge expenditure for constructing, operating, and maintaining the water infrastructure through user fees. However, the weaker sections of the society cannot afford to pay user fees, and the government has to subsidize the costs. In India, the poor cannot afford to pay for a water connection or buy water, and so they use surface water sources for their daily needs. Hence, surface water sources should be kept clean by controlling discharges and run-off into these sources. In many urban and rural areas of India, sewage and industrial effluents end up in waterbodies and spread a number of water-borne diseases.

The challenge for many municipalities is to find the money for building, operating, and maintaining water infrastructure. The Government of India has devised many schemes for building the required infrastructure, but the lack of proper cost accounting and collection of user fees has led to the improper maintenance of the infrastructure. Municipalities can conserve water and reduce their costs if they prevent losses in the water distribution and wastewater collection systems. A scheduled and well-planned maintenance programme is critical for the water infrastructure to effectively serve the people. Trained personnel and adequate funding are important, and all these can be accomplished through the collection of user fees.

Although India has passed laws on water cess (user fee), these laws are ineffective without modifications to reflect the costs required to build and maintain water infrastructure facilities. In addition, enforcement of the laws and charging of progressive user rates are critical for India to provide clean drinking water. Progressive user rates reflect the quantity of usage by consumers and their ability to pay. Consumers who conserve and recycle water can be encouraged by charging them lower user fees.

3.3 ENERGY RESOURCES

Energy is needed to maintain the quality of life and build economies. In areas where no other form of energy is available, people burn wood and other biomass for heating and cooking. In India, most poor people who do not have access to electricity and liquefied petroleum gas gather wood from forests. This has resulted in significant deforestation in some parts of the country. Most of the energy in India is generated by burning coal as there are large reserves of coal available in the country. Nuclear power accounts for only 5% of the energy, and there are plans to increase this source of energy. Hydropower is generated in northern India (in the Himalayan region and Punjab) and other parts of the country through large dams. India has taken an aggressive role in developing renewable energy resources, but this industry is still in its infancy. How the various sources of energy production affect climate change in India and other parts of the world is of concern to all human beings.

3.3.1 Use of Energy in Modern Society

In ancient times, humans planned their activity with the rising and setting of the sun. In many parts of rural India, this practice is still prevalent and the use of solar lamps or kerosene lamps is minimal. However, in modern cities, the availability of energy allows us to work and play for 24 h a day, 7 days a week. Energy is used from morning till night in homes and offices, from lighting our interiors and exteriors to provide comfort from the heat and cold. Many people in the modern society feel that if they have an electrical outlet at home, they will have energy on demand, but they have no idea of how the energy reaches the outlet. The average person rarely thinks about the impact of energy use on the environment, and many complain of blackouts and brownouts without considering what individuals can do to minimize these events.

Industries use a lot of energy for producing and transporting goods. When energy is cheap and plentiful, industry owners do not consider energy efficiency or conservation in their operations. However, rising costs of energy and the shortage of supply have forced many industries to pay serious attention to energy management and implementing various techniques to reduce energy uses. Many simple techniques are available for conserving energy, but an energy audit is required to determine how energy is being used. Energy use in transportation is heavily based on fossil fuels (diesel/petrol in cars, trains, and airplanes), which is a significant source of global warming.

Agriculture, a major source of employment in rural India, uses a lot of energy for water pumps, tractors, transport, and food processing. Solar applications for agricultural use are not yet widely implemented, although it should be relatively easy in India considering the large number of sunny days throughout the year in many parts of the country. The large amounts of biomass available in agricultural areas of the country can be converted to energy, thus saving fossil fuels for applications where alternative sources of energy are impractical (for example, in tractors and for transport to market) (Figure 3.3). Biomass can be converted to energy using gasifiers and biomethanation technologies. Animal and agricultural wastes (including corn and maize) can be converted to gaseous and liquid fuels for generating energy. Proper planning for the use of these wastes can return the agricultural areas of the country to energy self-sufficiency.

Figure 3.3 Types of biomass

Homes, schools, and other public facilities in rural areas can be run on alternative sources of energy if a concerted effort is made by the government and the industry. In many areas of India, solar energy is mobilized for those who are not connected to the electricity grid.

3.3.2 Coal: Our Abundant Source of Energy

India has large reserves of coal (bituminous, sub-bituminous, and lignite) located in all parts of the country (Figure 3.4). About 80% of India's energy needs are fulfilled by using coal as the fuel. The most efficient use of coal is to produce electricity at the mine mouth. The Neyveli Lignite mine in south India is one of the largest mine mouth power plants, where the mined coal is directly fed into boilers and converted to electricity for use in Tamil Nadu. Indian coal deposits have a large amount of ash, and the most effective use of such coal is to wash the

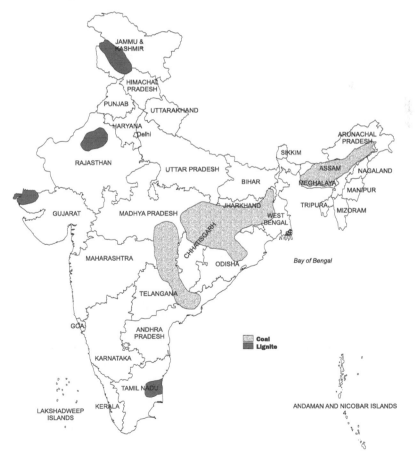

Figure 3.4 Major coal deposits in India

coal and remove the ash at the mine itself, which can be used for many construction projects in the mine. Transporting the coal with ash to a power plant, which then uses expensive pollution control equipment to clean the ash, is an expensive and inefficient use of resources.

Coal is burnt in boilers and converted to steam to run turbines that produce electricity. As a by-product of coal burning, ash is produced (both in the boiler grate and as fly ash, which escapes through the stack) and a large amount of carbon dioxide is generated. Research has proven that the emission of carbon dioxide leads to greenhouse effect and results in global warming and climate change. Clean coal technologies have been developed in Europe and the United States to improve the efficiency of coal plants by burning all the carbon in the coal. In certain "clean coal" technologies, coal is converted to a gas, which is then used to generate electricity (Figure 3.5). In theory, the carbon dioxide produced by burning coal can be sequestered underground to prevent it from causing global warming. The cost–benefit analysis and the availability of capital for such expensive technologies are major challenges in the adoption of these technologies.

Earlier the Indian Railways used a lot of coal for transporting goods and passengers. However, the air pollution caused by coal burning in locomotives prompted the railways to move towards diesel and electricity locomotives. Many poor people use coal in their houses and suffer from smoke and other pollutants from the inefficient burning of coal. Replacement of this fuel by biogas (from organic wastes) and other alternatives can improve the health of users and reduce global warming.

3.3.3 Nuclear Power

India produces only 5% of its energy from nuclear fuel. Nuclear power is an established technology used as a major source of power (up to 80%) in France, the United States (about 20%), and in many Western countries. Nuclear power does not produce GHG emissions, but it has the following major drawbacks:

- The health and safety issues are significant due to the potential radiation exposures to workers and to the public in the case of a radiation leak.
- Nuclear wastes remain harmful for more than 10,000 years and these have to be properly isolated from the environment.
- The capital costs are excessive compared to coal- or gas-fired power plants, and construction quality assurance has to be very rigorous.

CLEANER COAL COMBUSTION

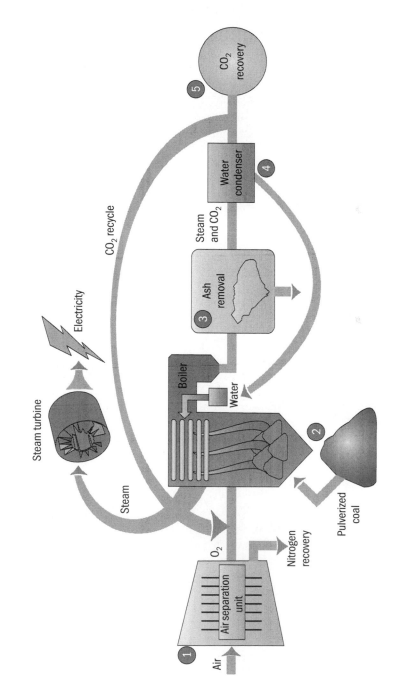

Figure 3.5 Cleaner coal combustion

- Plutonium is a by-product of nuclear waste reprocessing, and small amounts of plutonium can produce weapons of mass destruction.

Nuclear fuel is produced by processing uranium ore, and only 45 countries in the world are part of the Nuclear Suppliers Group (NSG); hence, getting assured supplies of nuclear fuel depends on the NSG agreeing to sell the fuel to a power producer. Earlier India could not expand its nuclear energy since it did not have sufficient uranium ore and the NSG countries did not allow it to buy additional ore or nuclear fuel because of the fear that nuclear fuel could be used for making weapons of mass destruction. However, India and the United States signed a nuclear deal with the concurrence of the NSG countries in 2008, under which India's nuclear power industry will be subject to inspection by the International Atomic Energy Agency.

Nuclear wastes consist of low-level and high-level radioactive wastes resulting from the spent nuclear fuel rods. Low-level wastes are radioactive from a few months to a few years and can be handled relatively easily in special landfills. However, high-level radioactive wastes are harmful for more than 10,000 years and are temporarily stored in water pools in the vicinity of nuclear reactors. Underground nuclear waste repositories have been developed in France, and many others are being prepared in the United States and Europe. Multiple engineered barriers in these repositories prevent the nuclear waste from contaminating water sources.

3.3.4 Hydropower

Hydropower has been produced for many generations by using the power of falling water in waterfalls. It was considered the most eco-friendly power source until ecologists found that hydroelectric dams negatively impact fish populations by disrupting their river habitats. India has large hydroelectric power plants in various parts of the country along major rivers. In addition, several hydroelectric power plants have been built and many more are under way along the Himalayan foothills, where the rivers originate.

The drawbacks of hydroelectric power are as follows:

- Large dams inundate large areas of land and displace people, as in the case of the Narmada dam (see case study in Chapter 12).
- Location of these power plants depends on the location of rivers, and expensive transmission lines are needed to bring the power to the population centres.
- When hydroelectric projects are constructed, ecological impacts occur on the fish and other organisms dependent on the river.

The major advantage of a hydroelectric power plant is that it does not produce GHGs and is relatively inexpensive compared to coal or nuclear power plants. In addition, it controls floods and produces multiple benefits for agriculture and the overall development of a region.

Hydroelectric projects have faced a lot of opposition because the benefits from the projects reach population centres far away from the people affected by the project. The inundated lands due to a large dam adversely impact farmers living on the land. Unless appropriate compensation is provided to these farmers and landowners, the project is delayed and sometimes public opposition is insurmountable. Adverse ecological impacts have forced the government to demolish many small, old dams in the United States. Many small hydroelectric projects are constructed in India to harness electricity for local use, and these have minimal adverse impacts.

3.3.5 Gas and Oil Power

Natural gas and petroleum or crude oil are depleting fossil fuels and have high demand for transportation. However, many power plants in India operate on gas and oil because these are the only fuels economically available at the plant location. Gas- and oil-powered plants also produce GHG emissions like coal-fired power plants.

Gas and oil resources are concentrated in a few areas of the world, especially in the Middle East where wars have been raging since 1948. Depending on oil and gas for a major portion of energy is not wise when these fuels adversely impact climate change. India has limited natural gas and oil resources, and a lot of money is required to explore and develop these resources. However, gas and oil power plants will exist in the foreseeable future, and alternative sources of energy must be found to replace these fuels.

The lack of reliable power in urban areas and the complete absence of energy in many rural areas of India force many people to use diesel engines for their power needs, which increases air pollution and GHGs. A priority for the Government of India is to develop alternative sources of energy and reduce the use of gas and oil for power production. Worldwide, the ongoing research and development on alternative sources of energy can economically replace the fossil fuel sources currently in use. In the future, it is hoped that policies and technologies will increase the role of nuclear power and renewable sources of energy such as sunlight, wind, and biomass.

3.3.6 Solar Energy

India is a leader in solar energy and is using solar energy in rural areas where it is the only source of energy. In 2009, India announced an ambitious programme for developing 20,000 MW of solar power by 2020. Researches in Japan, China, Europe, and the United States have focused on using solar energy for large-scale power production. India is blessed with a lot of sunshine during most of the year in most parts of the country. So it should take advantage of solar energy by passing legislations to encourage investment in solar power. Since 2009, incentives have been provided to the private sector to increase the use of solar power in rural and urban areas of the country and reduce global warming and decrease dependence on foreign sources of oil and gas.

Photovoltaic panels can be used in homes and businesses for their energy needs (Figure 3.6). Bringing down the cost of these panels and encouraging their use through tax credits will significantly increase the use of solar power in India. In California, tax policies and financial incentives have dramatically increased the use of solar power technologies by citizens and businesses. India has taken actions to encourage the development of alternative energies with the formation of the Ministry of New and Renewable Energy (MNRE). In addition, financial and tax policies have been introduced to encourage the wider use of solar energy in the country.

Energy use in commercial and government buildings can be converted to solar with encouragement from the government. Germany has been a leader in using solar technologies for replacing fuel sources that generate

Figure 3.6 Photovoltaic panels

GHGs. These technologies should be adopted in India and other countries to reduce GHGs and improve the environment. In addition, efficient transmission grids should be built to transmit solar energy from large solar power plants.

Solar energy is produced by the following two main types of devices:

1. Photovoltaic cells assembled in panels convert solar energy into electricity.
2. Solar panels in solar thermal systems heat water, which is used either directly or in steam turbines to produce large amounts of electricity.

China, Japan, and Korea are competing with Europe and the United States to become world leaders in solar energy technologies and for promoting the widespread use in cities and rural areas.

3.3.7 Wind Energy

Wind energy has been encouraged in India since the 1990s, and many wind farms have been developed in areas that have significant wind potential. Significant development of wind power technologies has been taking place in many countries, especially in Germany. Like the case of solar energy, storage and transmission technologies are critical for the growth of wind power in India. Wind energy is a green technology since it does not produce any GHG emissions. It is being produced in small quantities to serve local energy needs, and plans are under way to develop large wind farms that will produce large amounts of energy in areas with high-velocity winds, such as in coastal areas, and distribute the power to users.

Systematic studies are needed to identify the areas with consistently high wind velocities to maximize the production of electricity. Before setting wind farms, impacts on birds and the environment should also be carefully considered. Wind farms allow multiple use of land, and farming areas in rural communities can earn additional money by allowing wind energy production on their land.

Wind farms in the ocean and coastal areas have gained significant momentum in Europe and the United States. The Danish Government has developed wind farms off the coastline of Copenhagen and has policies that promote wind energy. In the United States, the state of Massachusetts is planning a large wind farm in the Atlantic Ocean off Nantucket Island. With a push for 10%–20% renewable energy in the United States, private investors are increasingly developing wind farms and solar energy.

3.3.8   Biomass Energy and Biofuels

The potential for biofuels and biomass energy has been exploited for the last 20 years, and significant potential exists for increasing energy production from these sources. A debate is going on about the use of corn and other sources of food for producing biofuels since it reduces food supply and the arable land needed to feed a growing population. Research on using grasses and discarded biomass materials (not sources of food) for biofuels is under way in many countries. Agricultural waste is a major source of biomass, which can be gasified for decentralized energy use in rural communities. These sources of green energy should be exploited further through more research and development, as well as appropriate fiscal and tax incentives. Biofuels replace fossil fuels and hence reduce GHG emissions (Figure 3.7).

India has been developing biomass gasifiers for the last 20 years, and many agricultural communities are using these to satisfy their energy needs. Gasifiers reduce GHG emissions and are eligible for carbon credits from the World Bank Carbon Prototype Fund. Biomass energy has significant potential to improve the lives of poor rural communities in India, Africa, China, and other developing countries and should be further developed by promoting research and development. In addition, carbon credits and other financial incentives should be fully explored to finance biomass energy projects in rural areas. A young entrepreneur in Bihar has developed small gasifiers using rice husk, and these can make agricultural communities independent of the power grid.[6] Small biogas plants using animal waste are being promoted in rural areas all over India.

In 2000, India's MoEFCC passed the Municipal Solid Waste Management Act, which forced many municipalities to convert all organic waste into energy through biomethanation plants.[7] In addition, many old landfills have been closed. Methane gas produced from waste decomposition is used and carbon credits are obtained, which makes the project economically viable.

3.3.9   Energy Conservation and Efficiency

The cheapest form of energy is the energy saved through conservation and efficiency improvements in the generation and transmission of electricity. There are several ways to improve the efficiency of energy generation:

---

[6] Husk Power Systems. Details available at <www.huskpowersystems.com/index.php>
[7] Ministry of Environment, Forests and Climate Change. Details available at <http://moef.nic.in/index.php>

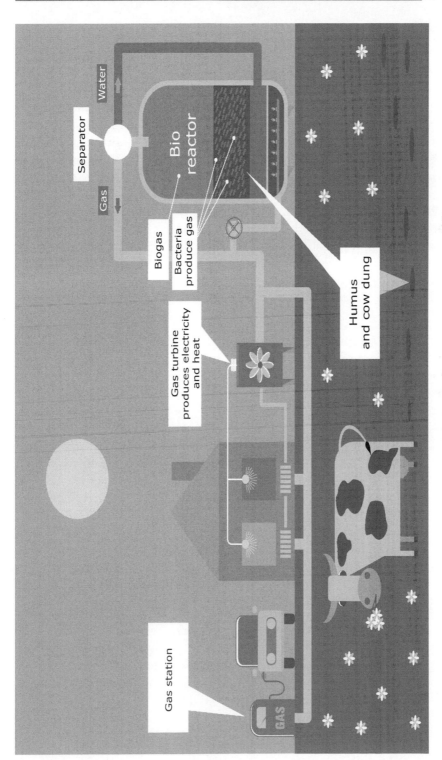

Figure 3.7 Utilization of biomass energy

- Minimizing heat losses and improving the combustion process
- Cogeneration by using the heat efficiently in an energy plant
- Proper operation and maintenance of power plants, as well as incorporating improved technologies in ageing power plants

Governments and electricity utilities have been promoting energy conservation measures for the past few decades. These include improved energy use in houses and commercial/government buildings and use of energy-efficient appliances and bulbs. Awareness among energy users and financial incentives for promoting conservation are essential for successfully implementing these measures. Teachers have a wonderful opportunity to impart the importance of energy conservation in classrooms and school buildings on their students. This will have a huge impact on the community since children can ask their parents to implement energy efficiency in their houses and offices.

3.3.10   Global Warming and Energy

Global warming and energy use are closely linked. Energy is produced from various fuel sources, some of which cause GHG emissions. There is a global awareness that the use of fossil fuels should be reduced and more energy should be produced from renewable sources such as wind, biomass, and sunlight. Since the Intergovernmental Panel on Climate Change was set up by the United Nations, global agreements have been reached to reduce GHG emissions from fossil fuels and promote the growth of forests to absorb GHG. The use of wind and solar energy has significantly increased worldwide, and a lot of research is being conducted on bio-energy. India is planning to increase its nuclear energy along with various sources of renewable energy. The MNRE is conducting extensive research and promoting the development of alternative energy resources. All these efforts are reducing global warming, while producing the energy required by the growing population. In addition, energy efficiency has received a lot of attention from various sectors of the economy.

3.4   MINERAL RESOURCES

India is endowed with a lot of mineral resources, which are exploited for producing energy, constructing infrastructure, and supporting a modern society. The data of mineral resources/reserves, as per the United Nations Framework of Classification for Fossil Energy and Minerals Reserves and Resources 2009, are compiled and maintained by the Indian Bureau of Mines. The data cover detailed information on various items and deposit-wise mineral inventory. Updated at regular intervals, the data are published annually in the form of the *Indian Minerals Yearbook*. This yearbook covers mineral prospects/deposits/mines in freehold and leasehold areas, their status, infrastructure, geology and exploration, ore characteristics, estimated reserve/resource, details of feasibility, and details of mining along with production data. The data are sourced from various exploration agencies, including Geological Survey of India, state Directorate General and Mines (DGM), and public and private sector mining organizations. This report currently comprises 16,000 deposits, of which 8000 are in freehold areas, 800 are in public sector leasehold, 7100 are in private leasehold, and 100 in partly leasehold.

3.4.1 Mineral Production

India has significantly large resources of iron ore, bauxite, chromium, manganese ore, baryte, rare earths, and mineral salts. In India, minerals are broadly classified into minor minerals and major minerals. At present, there are more than 3700 active major mines in India, employing more than 5 lakh people. India produced 90 minerals in 2012/13, which included 4 fuel, 11 metallic, 52 non-metallic, and 23 minor minerals. The value of mineral production in 2012/13 was ₹285,761 crore (approximately $44.65 billion), which was about 2.4% of the gross domestic product. Of the total production, fuel minerals account for 64%, metallic minerals 15%, non-metallic minerals 3%, and minor minerals 18%.

3.4.2 Mineral and Mining Sector Legislations in India

3.4.2.1 Overview

India's mineral and mining sector operates under a federal structure, wherein the Central Government formulates the legislation for all minerals, except minor minerals, and the state governments formulate legislations for minor minerals. The Government of India permits 100% foreign direct investment in exploration, mining, mineral processing, and metallurgy through the automatic route, by way of equity participation in a company incorporated in India, for all non-fuel and non-atomic minerals. India has written legal and constitutional framework to manage the mineral sector. The National Mineral Policy provides the direction for the mineral sector. The central and state governments are responsible for the management of the mining sector. The state governments own minerals occurring onshore. In the case of offshore areas, the ownership of minerals rests exclusively with the Central Government. The Constitution of India bestows power to the parliament to enact legislations relating to mining, and the states are bound by the central legislations.

3.4.2.2   Legislations

- The *Mines and Mineral (Development and Regulation) (MMDR) Act, 1957* is the central legislation for regulating mining operations. The act enables all state governments to exercise their powers within a uniform national framework. The state governments, as owners of onshore minerals, grant mineral concessions and collect royalty, dead rent, and fees as per provisions of the act.

- The *Mineral Concession Rules (MCR), 1960* define the process of granting mineral concessions as per Section 13 of the MMDR Act. The rules lay down the process and timelines for granting concessions, disposal and refusal of applications, and the basic conduct of accounts, registers, and information reports.

- The *Mineral Conservation and Development Rules (MCDR), 1988* prescribe guidelines for the conservation and development of minerals as per Section 18 of the MMDR Act. The rules prescribe procedures for carrying out prospecting and mining operations and the general requirements relating to the preparation of mining and prospecting plans and filing of notices and returns. The rules also cover guidelines for the protection of the environment.

- The *Mines Act, 1952* prescribes the laws relating to labour safety in mines and regulations for carrying out mining operations and management of mines. The act lays down rules for the health and safety of the people employed in mines and regulates their working conditions. It also contains provisions relating to inspection of mines and reporting procedures.

- The *Mines Rules, 1955* define the framework for the medical examination of persons employed or to be employed in mines, basic health and sanitation provisions, and welfare amenities for miners and their families.

- The *State Minor Mineral Concession Rules* are prescribed by various state governments for granting concessions on minerals classified as minor minerals under the MMDR Act.

- The *Offshore Areas Mineral (Development and Regulation) Act, 2002* regulates the mining and development of minerals in offshore areas. The act provides for the development and regulation of mineral resources in territorial waters, continental shelf, exclusive economic zones, and other maritime zones of India. It empowers the Central Government to grant mineral concessions for offshore areas and collect royalty. The Indian Bureau of Mines is notified as the administrative authority.

- The *Offshore Areas Mineral Concession Rules, 2006* lay the process for the grant and renewal of reconnaissance permits, exploration licenses, and production leases as per Section 35 of the Offshore Areas Mineral (Development and Regulation) Act, 2002. The rules prescribe measures for protecting the marine environment and safety measures to be followed in leased areas. The rules also define the operational guidelines for each concession granted under the 2002 act.

3.4.2.3 Recent developments

The MMDR Act, 1957 was amended through the MMDR Amendment Act, 2015. The amendment that came into force on 12 January 2015 has ushered in the regime of transparent and non-discretionary grant of mineral concessions. The major features of the amended act are as follows:

- Mining leases will now be granted for a term of 50 years.
- The mineral concessions will now be granted through the auction process and will not be renewed after the expiry of the concession.
- The Central Government will prescribe the terms and conditions for the grant of mineral concessions through competitive bidding.
- Reconnaissance permits will henceforth be granted on non-exclusive basis.
- The Central Government has the authority to reserve mines for specific end uses at its discretion.
- A district mineral foundation is to be set up in each mineral bearing district for local area development.
- National Mineral Exploration Trust is to be set up for regional and detailed exploration in the country.
- The Government of India is in the process of simplifying and updating the subordinate legislations relating to the mineral and mining sector in India, which include necessary amendments to MCR 1960 and MCDR 1988. As part of this initiative, the government has notified the following rules for the implementation of the MMDR Amendment Act, 2015.
  - The *Minerals (Evidence of Mineral Contents) Rules, 2015* prescribe procedures for conducting exploration to determine mineral content so that mineral blocks can be taken up for auction of mineral concessions.

- The *Mineral (Auction) Rules, 2015* detail the process to be followed for auction with respect to grant of mineral concessions.

- The *Mineral (Non-exclusive Reconnaissance Permits) Rules, 2015* detail the process to be followed for the grant of non-exclusive reconnaissance permits.

- The *National Mineral Exploration Trust Rules, 2015* detail the objectives, functions, and operations of the National Mineral Exploration Trust.

3.5 FOREST RESOURCES

A forest is a complex ecosystem predominantly composed of trees and shrubs and is usually a closed canopy. Forests are storehouses of a large variety of life forms such as plants, mammals, birds, insects, and reptiles. Also, forests have abundant microorganisms and fungi, which decompose dead organic matter and enrich the soil. Nearly 4 billion hectares of forest cover the earth's surface, which is roughly 30% of the total land area.

The forest ecosystem has two components: non-living (abiotic) and living (biotic). Climate and soil type are part of the non-living component, and the living component includes plants, animals, and other life forms. Plants include trees, shrubs, climbers, grasses, and herbs. Depending on the physical, geographical, climatic, and ecological factors, there are different types of forests, such as evergreen forest (mainly composed of evergreen species, that is, species having leaves throughout the year) and deciduous forest (mainly composed of deciduous species, that is, species whose leaves fall during particular months of the year). Each forest type forms a habitat for a specific community of animals that are adapted to it.

The term forest implies "natural vegetation" of the area, existing for thousands of years and supporting a variety of biodiversity, forming a complex ecosystem. Plantation is different from a natural forest as the planted species are often of the same type and do not support a variety of natural biodiversity.

3.5.1 Forest Cover in India

As per the *India State of Forest Report, 2009*, the total forest cover of the country (2007 assessment) is 690,899 km², which is 21.02% of the geographic area of the country. Of this, 83,510 km² (12%) is very dense forest (VDF), 297,087 km² (43%) is moderately dense forest (MDF), and 269,699 km² (39%) is open forest cover (OF). The scrub accounts for 41,525 km² (6%).

Forest cover includes all lands that have a tree canopy density of 10% and above with an area of 1 ha or more. VDFs have tree canopy density of 70% and above. MDFs have all lands with tree cover of canopy density between 40% and 70%. OFs have all lands with tree cover of canopy density between 10% and 40%. Scrub includes degraded forest lands with canopy density less than 10%. Non-forest is any area that is not included in the aforementioned classes.

Areas under VDF, MDF, and OF also include mangrove cover of the corresponding density class.

3.5.2 Tree Cover in India

All areas with an extent of more than 1 ha and with 10% or more tree canopy density are included under forest cover. However, there are many small patches of trees less than 1 ha in extent, such as trees in small-scale plantations, woodlots, or scattered trees in farms, homesteads and urban areas, or trees along linear features such as roads, canals, and bunds, which are not captured by satellite sensors owing to technological limitations. All such patches are included under "tree cover".

As per the *India State of Forest Report, 2009*, the tree cover in the country was 92,769 km^2 in 2007, which is 2.82% of the total geographical area. Tree cover constitutes the largest area in Maharashtra, followed by Gujarat, Rajasthan, and Uttar Pradesh.

3.5.3 State of Forest Resources

Forests include the areas under the forestry land-use category and some areas recorded as forests and not under tree cover, such as rocky areas, deserts, and mountain ranges. The recorded forests, inclusive of these categories, extend over 76.962 million ha, which is 23.41% of the geographical area of the country. The forest cover in the country is 67.70 million ha, which is 20.60% of the geographical area of the country.

The country's forests have been grouped into five major categories and 16 types according to biophysical criteria. The distribution of these groups includes 38.20% subtropical dry deciduous, 30.30% tropical moist deciduous, 6.7% subtropical thorn, and 5.8% tropical wet evergreen forests. Other categories include subtropical pine (5%), tropical semi-evergreen forests (2.5%), and other smaller categories. Temperate and alpine areas cover about 10% of the forest areas in the Himalayan region.

3.5.4 Forestry and Livelihoods

Forests and forest products have been recognized as multipurpose resources with the potential of providing livelihoods to a substantial part of the population. Constitutional provisions empower local panchayats with the rights to use non-wood forest products. The strategy for forest management focuses on empowerment of community institutions in managing and deriving livelihoods from forests. The Government of India has promulgated a law recognizing the rights of forest dwellers in forest lands wherein ownership rights are documented and the rights of communities are recognized for the common use of forests for their livelihood practices conforming to the principles of sustainability of forests. These provisions enhance the stake of forest dwellers and fringe populations in the development of forests. The outcome of this new law will be visible after a few years.

Similarly, with the realization that the existing forest areas alone cannot cater to all the livelihood needs of the country, sizeable areas of non-forest lands used earlier by rural communities as common property resources need to be revived and regenerated. The Constitution of India puts the responsibility of social forestry on local panchayats. The revival and management of such village common lands are promising for the sustainable management of natural resources.

3.6 LAND RESOURCES

India has 1.3 million square miles of land, protruding into the Indian Ocean and located between the Bay of Bengal on the east and the Arabian Sea on the west. The types of land include 43% in the plains, 30% in the mountains, and 27% in plateaus. In spite of the sufficient accessibility to land resources, the pressure of population in the country is excessive and makes space for both food production and real estate market at a premium. Increased urbanization and industrialization have put a lot of pressure on agricultural land.

The Government of India is reorganizing its efforts in land management to utilize land resources effectively. Agricultural land occupies 56.8% of the total land area, and about 70% of the population lives in these rural areas. Land resources in India include vast barren lands in Rajasthan, parts of Leh, and Jammu. Housing demand is increasing owing to the growing middle class and large tracts of agricultural land have been taken over by the real estate industry.

4

Biodiversity

4.1 WHAT IS BIODIVERSITY?

The concept of diversity in living species, big and small—elephant or frog, potato or butterfly—is called "biodiversity". The term biodiversity was first coined in 1985 and was subsequently made popular in 1988 by the famed entomologist E O Wilson (Wilson 1988). Biodiversity is defined as the "variation of life at all levels of biological organization" (Gaston and Spicer 2004). This encompasses the variation at the genetic and ecosystem levels. Currently, preserving biodiversity and protecting rare and endangered plants and animals throughout the world have become increasingly important in environmental conservation efforts by the United Nations and other global organizations. Biodiversity influences the web of life, which depends on the interconnectedness of ecosystems and species. Without biodiversity, the web of life breaks down and species become extinct.

In 2006, the UN General Assembly designated the year 2010 as the International Year of Biodiversity. This coincided with the 2010 biodiversity target that was first adopted by the European Union (EU) at the EU Summit 2001 to halt the decline of biodiversity by 2010.[1] One year later, the Conference of the Parties, the governing body of the Convention on Biological Diversity (a legally binding international treaty), adopted the Strategic Plan for the Convention in Decision VI/26. This decision stated that "Parties commit themselves ... to achieve by 2010 a significant reduction of the current rate of biodiversity loss at the global, regional and national level as a contribution to poverty alleviation and to the benefit of all life on earth".[2] Therefore, biodiversity has become the focus of environmental efforts at the global level in a relatively short time.

---

[1] Details available at <www.countdown2010.net/biodiversity/the-2010-biodiversity-target>

[2] COP 6 Decision VI/26, Annex B

4.2 BIODIVERSITY IN INDIA

As the seventh largest country in the world, India contains a variety of species of plants and animals distributed among its different ecosystems and geographical regions.[3] India has four biologically distinct zones or biomes: the Himalayas in the north, the Ganges river plains extending from east to west, the Deccan Plateau in the south, and the Andaman and Nicobar Islands and Lakshadweep Islands. These regions contain three main types of ecosystems: wetlands, forests, and marine habitats.

4.2.1 Wetlands

A wetland is a land area saturated with water, permanently or seasonally, and takes on the characteristics of a distinct ecosystem. Wetlands are important for biodiversity because they support a wide range of plants and animals. They are effective for flood control, water purification, and shoreline stability and act as carbon sinks. Wetlands cover about 20% of the total geographical area of India.[4] Most of the wetlands in India are under paddy cultivation and, therefore, provide economic benefit to the local people. Two especially important sites are significant waterfowl habitats: Chilika Lake in Odisha and Keoladeo National Park in Bharatpur. India has nine major wetlands covering many regions: the reservoirs of the Deccan Plateau/lagoons of the south-west coast, the saline expanses in Rajasthan/Gujarat, the freshwater lakes and reservoirs in Rajasthan/Gujarat, Chilika Lake and the eastern delta wetlands, the freshwater marshes of the Ganges plain, the floodplain of the Brahmaputra river, the marshes and swamps of the north-east, the mountain lakes and rivers of Kashmir and the Himalayan region, and the mangrove forests in the Andaman and Nicobar Islands (Scott 1989).

4.2.2 Forests

In India, forests range from the tropical evergreen forests of the Andaman and Nicobar Islands to the dry alpine scrub in the mountainous regions of the Himalayas. Between these two extremes are semi-evergreen forests, deciduous monsoon forests, thorn forests, subtropical pine forests in the foothills of the Himalayas, and finally the temperate forests of the mountainous regions (Lal 1989). Some of the forest regions, particularly the monsoon forests of the Western Ghats, produce trees that have

---

[3] Details available at <http://indiabiodiversity.org/maps>
[4] Ministry of Environment, Forests and Climate Change. Conservation of Wetlands in India: A Profile. Details available at http://envfor.nic.in/divisions/csurv/WWD_Booklet.pdf

significant commercial value, such as the Indian rosewood. Many of the evergreen forests are currently degrading into semi-evergreen forests as a result of human interference, such as clear cutting for agricultural purposes; this leads to loss of biodiversity as the ecosystem changes.

4.2.3 Marine Habitats

The coastal waters that surround most of the Indian subcontinent contain rich fishing areas of great economic importance. Reef areas, such as the Gulf of Kutch and the Gulf of Mannar, are currently exploited for coral, coral debris, ornamental shells, pearl oysters, sand, and spiny lobster fishing. The mangroves protect the shore from erosion and flooding and support fish nurseries and prawn farming (Scott 1989).

These areas must be protected to prevent overexploitation and subsequent reduction in biodiversity. For example, all the five species of marine turtles native to the Indian coastal waters are currently declining in number due to direct human predation and accidental catch during commercial fishing.[5] If the biodiversity of marine areas is not protected, other species of marine flora and fauna will face the same fate. Therefore, ways must be found to protect biodiversity, while allowing local populations to earn their living.

4.3 SPECIES DIVERSITY

Dozens of plant and animal species are endemic to India. They are unique to a particular geographic region and are not found elsewhere. North-eastern India, the Western Ghats, the north-eastern and north-western Himalayas, and the Andaman and Nicobar Islands are all particularly rich in endemic species of plants. Endemic species are of particular importance to conservationists because their habitat is generally small and the entire species may be threatened by factors such as human development or climate change.

India has 44 endemic vertebrate species. According to the International Union for Conservation of Nature, conservation of the following four species is especially significant: lion tailed macaque, Nilgiri langur, brown palm civet, and Nilgiri tahr.

India has 55 endemic bird species, which are particularly concentrated in areas of high rainfall. The highest number of endemic species in India can be found among reptiles (187) and amphibians (110).

Protecting endemic species of plants and animals is especially important because the loss of these species will irrevocably affect global

---

[5] Details available at <www.iucn.org/?uNewsID=833>

biodiversity as they are not found elsewhere. In addition, endemic species are developed by specific evolutionary pressures and are often uniquely adapted to their environment. Therefore, any changes in the local environment may drastically affect the ecosystem as a whole.

4.4 THREATENED SPECIES IN INDIA

India contains globally important populations of some of Asia's rarest and most important animals. These include the Bengal fox, Asiatic cheetah, marbled cat, Asiatic lion, Indian elephant, Asiatic wild ass, and the Indian rhinoceros (Figure 4.1). Some of these, such as the Indian elephant, are important symbols of the country and are threatened by human encroachment, habitat destruction, and poaching (Choudhury, Lahiri Choudhury, *et al.* 2008). Enforcing measures to protect these species is critical owing to their cultural and biological significance.

Figure 4.1 Asiatic cheetah and Indian elephant are two of Asia's rarest species

4.5 NATIONAL POLICY FOR PROTECTION OF BIODIVERSITY

In India, individual states oversee the task of protecting species and biodiversity, with minimal intervention of the Central Government. India adopted the National Policy for Wildlife Conservation in 1970. This was soon followed by the Wildlife (Protection) Act, 1972. This act set up state wildlife advisory boards and created wildlife wardens within state forestry departments. This created a dedicated cadre of individuals with primary concern of wildlife protection. The management of protected areas varies from region to region. For example, in the Nilgiri area of Tamil Nadu, no human intervention is allowed in protected areas, whereas in Kerala, efforts have been made to balance commercial endeavours with protection goals.

The aforementioned legislations have encouraged the growth of protected areas and national parks throughout India. In the Western Ghats, for example, seven national parks have been established to protect the large number of species indigenous to the region. In addition, India adopted two species-specific acts of importance. One is the Project Tiger, adopted in 1973 by the Central Government in association with the World Wildlife Fund and dedicated to protect the Bengal tiger. The other is the Crocodile Breeding and Management Project of 1975, dedicated to the protection of native crocodile populations (Saharia 1982).

4.6   INTERNATIONAL ACCORDS HELPING INDIA SUPPORT BIODIVERSITY PROTECTION

In 1976, India became a signatory to the Convention of International Trade in Endangered Species. This is a commitment by all signatory nations to prohibit trade in wild plants and animals, thereby eliminating the market for endangered animal or plant specimens. In 1977, India became a signatory to the United Nations' World Heritage Convention, which led to five national parks being designated as "of outstanding universal value". In addition, India signed the Convention on Biological Diversity in 1992 and adopted it in 1994. This convention provides a framework for the sustainable management of natural resources. In 1982, India also became a signatory to the Ramsar Convention for the protection of wetlands. This created six areas of protected wetlands across India. From local to international levels, India has taken seriously the task of protecting biodiversity.

4.7   INVASIVE OR ALIEN SPECIES

Alien species are plants and animals from other countries or continents that have invaded India, spreading quickly and displacing native species. Whether introduced deliberately or by accident, invasive species are second only to habitat loss as the cause of deteriorating biodiversity. India has only 2.4% of the world's land area but 8% of global biodiversity. Unfortunately, India also has suffered a profusion of invasive species. *Invasive Alien Flora of India* lists 173 troublesome species that threaten native ecosystems.

Three plant species are particularly damaging to India's biodiversity: the Siam weed, *Lantana*, and water hyacinth. Siam weed (*Chromolaena odorata*) can take over large tracts of land. *Lantana* (*Lantana camara*) or wild sage is another invasive plant that has wreaked havoc on endemic flora. Introduced in India as an ornamental plant around 1810, it spread quickly. *Lantana* suppresses native species with its allelopathic effect: It secretes a chemical that inhibits the growth of other plants around it, making native plants unable to compete.

Water hyacinth (*Eichhornia crassipes*) is another ornamental plant introduced in India during 1914–16. It is an aggressive colonizer that flourishes in still freshwater bodies in South India, including the backwaters of Kerala. Local species cannot compete, and their survival is threatened.

India and Southeast Asia have been the victims of invasive species, but they have also exported some alien species to other countries. In India, the mongoose preys on snakes. When settlers arrived on the Hawaiian Islands and West Indies, they unintentionally introduced rats to the local ecosystem. In order to eradicate the rats, they introduced mongoose from India. However, in addition to hunting rats, the mongoose preyed on endemic ground-nesting birds and their eggs. This decimated some local bird species, causing their extinction.

Introduced species do not always compete with native species. The potato originated in South America, but it is widely grown in India without pushing out native plants. Tea (*Camellia sinensis*) came from China, yet tea industry is the major contributor to the economy of India and Sri Lanka. Tea crops can flourish alongside local flora without challenging their survival, thereby maintaining India's biodiversity while allowing local people to earn their livelihoods.

5

Population and its Impact

5.1 WORLD POPULATION

India is growing! China and the United States are also growing! These countries are growing not in terms of area but in terms of population. They are the world's three largest populated countries. The population of these countries is growing for different reasons. India has a high birth rate, with each couple giving birth to an average of 2.59 babies. China's birth rate is only 1.58 babies per couple, but China has a "bulge" in the number of mothers of child-bearing age. In the United States, immigration adds to the population even if the birth rate is relatively low at 1.89 babies per couple.

The growth in rest of the planet is staggering, the totals are also staggering. Can the world sustain this population growth? Currently the world has more than 7 billion people which according to the United Nations was achieved on 31 October 2011.[1] The total reached 6 billion shortly before 2000, adding over a billion in a person's lifetime. The world's population has doubled since 1972.

Since October 2011, the monthly growth of world population has been about 7.7 million, which is equivalent of adding another Chennai 12 times in a year. The daily growth is close to 260,000. Across the world, that is more than two more births than deaths every second.

5.2 INDIA'S POPULATION

India is the second most populous country on earth, with 17% of the world's population dependent on just more than 9% of the arable land. Remove the 200 million people of Uttar Pradesh from the total, and India would still be the second most populous country on earth. Yet if Uttar Pradesh were a separate country, it would be the fifth largest, polluted country with more people than either Pakistan or Bangladesh.

India's population passed the 1 billion mark in May 2000, and Pakistan and Bangladesh have kept pace. Both countries have tripled or

---

[1] Details available at www.un.org/popin

quadrupled in population and density since partition. Bangladesh is the most densely populated country in the world, with more than 950 people per square kilometre.

According to 2001 census, India's population was 21% greater than the previous census 10 years earlier and had nearly doubled since 1971. In 2011 census, the total population was 1.21 billion. If this increase continues, India's population could increase to 2 billion by 2050. Birth rate in India is decreasing, but a decrease in birth rate is not the same as levelling or decrease in population. At the current growth rate of 1.5% per year, India's population will double during a person's lifetime. India could overtake China as the world's most populous country as early as 2025. China is committed to maintaining its population. Since 1979, the Chinese Government had allowed each family to have only one child. From 2015, this rule is being phased out.

India, however, cannot follow the same path of one-child policy. Compared to China, India is incredibly diverse in language, religion, economy, and culture. Diverse backgrounds produce divergent viewpoints about population. Most importantly, India is a democracy, and any government population policy must have the understanding and support of the people.

Projections of future population of India are very uncertain. Some countries have undergone a "demographic transition", changing from large families and a high growth rate to small families and a low growth rate, or even a stable population. Governments can encourage such a transition with policies and programmes that include education, economic incentives, and family planning services.

Some changes have taken place in India, and the growth rate has slowed in the recent years. Better health care and education makes life more stable. As people feel more secure, both socially and economically, family size drops. Will the population growth rate in India continue to drop, by planning or by circumstance?

Total population is important, but a focus on the total overlooks other important trends, including internal migration from villages to cities. People concentrate in cities for work and other opportunities. On average, now more people live in cities and towns than in the countryside throughout the world. Since 2000, the world average for urban dwellers passed 50%. India is well below that; only 28% of people in India live in towns and cities, and the rest, mostly farmers, live in rural areas. As Indians move off the farms, the cities will swell. In 1931, there were fewer than 35 million urban dwellers in India. By 2020, Mumbai could be the largest city in the world with 35 million people. Kolkata and Delhi would not be far behind at 30 million each.

In 2001 census, over a third of the India's population was under 14 years and only 7.5% was over 60 years. Even if individual families are kept small, the many Indians in the child-bearing age will make the population to grow. The current population has strained the country's systems, natural or man-made. Rivers have either dried or become polluted, and wells have gone dry; roads and railways have become overcrowded. For its 17% of the world's population, India has only 4% of the world's freshwater. If India wants to be a world leader, it should meet the challenges of educating the youth and feeding the nation. Accommodating a growing population will not be easy.

5.3 POPULATION CONTROL

Clearly, the world's current growth in population is unsustainable. Indian leaders had recognized overpopulation as a concern as early as in the 1930s, often spurred by food shortages, even starvation. In 1930, the Maharajah of Mysore opened a birth control clinic, the first in the world sponsored by a government. The awareness continued to India's independence and beyond. Jawaharlal Nehru commented that food production was increasing along with the number of children but wished for fewer children. By the 1950s, population increase was dramatic, caused by a high birth rate coupled with people living longer.

In 1952, India was the first country to have a plan to limit population growth; other countries used India as a model. Despite an official media campaign for one or two children per family, the birth rate rose in the succeeding decade. By the mid-1960s, the Indira Gandhi government stepped up control efforts, shifting to specific measures. After successive droughts, leaders blamed excess population for more food shortages. By 1968, the goal was to reduce the birth rate by 45% in 10 years.

National efforts promoted voluntary sterilization, primarily vasectomies for men. Incentives were offered: cash or a transistor radio. For the poor, a payment equal to several weeks' salary was hard to resist. From 1961 to 1971, India added 100 million people, and the government responded. Besides declaring a state of emergency, the Indira Gandhi government adopted a more coercive population policy. Compulsory sterilization was mentioned. With the centre's backing, the states used forceful methods.

Some of the methods were harsh, and particular groups, including tribals, were targeted. The opposition hardened its stance. The government overreached, resulting in resentment and protests. September and October were the key months in 1976 in the outcome of the administration's actions. Promotion of sterilization, though pushy, was still supposedly voluntary. In September, sterilization became compulsory. Some groups were singled out, and protests escalated. In October, a protest escalated in Muzaffarnagar in western Uttar Pradesh. It turned violent as the mob torched a clinic and the police fired shots. Many died. In his well-researched 2007 book *India after Gandhi*, Ramachandra Guha called the calamity the "Turkman Gate of family planning", recalling a deadly reaction to slum clearing in Delhi earlier in 1976. At first, news of the incident was suppressed, but outrage followed and the government had to back down. After the state of emergency was lifted and new elections were held in early 1977, the programmes were either limited or closed.

There will always be opposition. Because of the diversity of the Indian people, there are sharp differences in opinion within the country: urban versus rural, north versus south, east versus west, and based on ethnic groups as well as educational and economic levels. Some religions reject all methods of controlling population growth. Even individuals who agree that overall population limits are necessary will often argue that their group or interest is special and should not have controls. Initiating a programme does not mean that it is followed. Without individual and public support, any plan is ineffective.

Countrywide, there are still population control programmes. India adopted the National Population Policy in 1976, which included overall well-being. The experts reasoned that healthy, educated individuals would naturally limit family size. In 2000, an updated policy set the goal of a stable population by 2045.

Population control advocates taking the help of media to promote the message. Doordarshan, the country's Central Government founded public service broadcaster, presented shows on social issues. *Hum Log* (We the People), which showed middle class families, quickly became popular in 1984 and 1985. Sensitive subjects fit easily into soap operas, and *Hum Log* had them all: women's status, family planning, and more. Almost as popular was *Humraahi* (Come Along with Me) in 1992, despite competition from cable television. The set-up was the same, entertainment with social content, and popular singers and actors boosted ratings as they repeated the theme of the show. Telenovelas showed that media could change people's thinking. Indian television likely did the same.

During the same years as Indian politicians worried about population, leaders in China fostered growth. Chinese Communist ruler Mao Tse-tung thought more children would make China more powerful. Despite a famine during the "Great Leap Forward", covered up until after Mao's death, and the upheaval of the Cultural Revolution in the 1960s, China's population boomed.

Even when Mao was in power, officials began adjusting his policies. After his death in 1976, the country again promoted birth control and reproductive education. Later in the same decade, China began the strategy for which it is best known: the one-child policy. Signboards in cities and villages publicized the message. At that time, China had about a quarter of the world's population living on just under 9% of global arable land. China's population was young, two-thirds of the people were under 30, and the government was intent on economic reforms to lift the standard of living. As per Chinese Government, over 25 years, the policy limited China's population by 400 million. Chinese culture, however, favours male offspring. The numbers confirm the bias: A normal ratio is 105 boys born for every 100 girls. In China after 2000, the ratio was about 85 girls to 100 boys. In 2015, the government started phasing out this one-child policy.

There were a few exceptions to the one-child policy: ethnic minorities were allowed larger families and the quota was not strictly enforced in rural areas. China has eased the policy, and the one child per family was never absolute. In November 2013, the government allowed an additional exception: if either parent in a new family was from a one-child family, the couple is allowed to have two children. Still, for most of the population, it was one child per couple, well below the replacement rate of 2.1 per family. No other country had been so committed to reducing birth rate so successfully for so long.

Despite the low birth rate, below replacement, the population of China is growing, which is a seeming contradiction. This is because the baby boomers of the Mao years are now having children, and the boomer's large numbers overpower the low birth rate.

The Chinese preference for male offspring is matched in India. The numbers confirm the preference. As of 2011, there were 940 women for every 1000 men, up from 933 in 2001. These results, however, cannot be accepted. Fewer females mean fewer brides, and men may seek a mate from a very different background, region, or language.

Outside of India and China, birth rates are dropping in most countries, but the population is still growing. If the birth rate is higher than the death rate or there is significant immigration, population will increase. Beyond India and China, international issues are best looked at through the United Nations, specifically its Department of Economic and Social Affairs (DESA). The sessions held by DESA have followed world thinking about population stabilization, first mostly by emphasis on control of growth rates and later shifting to education and women's rights.

5.4 MALTHUSIAN CONCEPTS ON POPULATION

Population controversies can be best known by reading Reverend Thomas Malthus, an English clergyman and influential thinker about population. His *Essay on Population* (1798) predicted that the world's population would grow faster than its food supply, leading to undernourishment and then famine. Half a century after Malthus's essay, a million people died in the Irish potato famine. The followers of Malthus thought they had their proof, but then the world changed around them. World population now is 10 times what it was when he wrote the essay, and despite regional starvation and devastating famines, most people have adequate food. Rich croplands, particularly in the Americas, were farmed for the first time in the century after his prediction. Improved farming methods and the use of synthetic fertilizers gave far better crop yields. After the Second World War, new strains of major grains (wheat, maize, and rice) also increased yields as part of the Green Revolution. Punjab launched the Green Revolution in India and other states followed. Two centuries after Malthus, his prediction of catastrophe has not come true.

India has had famines before, during, and after the British rule. Regardless of the regime, failure in monsoon signals trouble. Oral tradition and ancient literature extend the history of India's famines as far back as 269 BC. The "Great Famine" lasted from 1876 to 1878. Estimates of the death toll ranged upwards from 5 million to more than 25 million. Bengal had a famine from 1942 to 1945, as did Bihar in 1966, though on a much smaller scale. After the Great Famine, spotting warning signs of food shortage led to better organization in India to head off a famine in the making.

In the shadow of famine is malnutrition. Polluted water and living conditions weaken people, and disease claims the undernourished before they starve. Lurking behind malnutrition is food insecurity, not knowing if your next meal will be adequate. One method India uses to secure nutrition is the "Midday Meal Scheme", providing free lunches in schools across the country, the largest such effort in the world. Besides improved health, better school attendance is a benefit for the children served, over 100 million. A tragic mishap blemished the programme in July 2013 when 23 students at a school in Bihar died from poisoning, likely from a pesticide that accidentally contaminated the cooking oil.

Even if a person, particularly a child, has enough calories, they may not be getting enough other dietary necessities such as vitamins, minerals, and proteins. Poverty makes it tougher. According to the Population Reference Bureau, 76% of Indians live below the poverty line defined as an income of $2 a day, which is about ₹120. Only 26% is the official government estimate which is still 300 million, more than the number of people living in any country after the top three. While many of the poor are in villages, many also live in cities, increasingly joined there by migrants from the country. Experts agree that the percentage of Indians living below the poverty line is decreasing. Because the population is growing, the number of poor remains high.

Will India be the country that proves that Malthus was right? Famously, in his influential 1968 book, *The Population Bomb*, Paul R Ehrlich invoked Malthus when he predicted mass starvation in India in the 1980s. It did not happen in that decade or since. Regardless, the neo-Malthusian pessimists assert that population disaster looms.

5.5 POPULATION GROWTH AND ITS IMPACT ON THE ENVIRONMENT

Every person, rich or poor, has an impact on the environment. Everyone needs the basics: food and water, air to breathe, and a place to live. Beyond the basics, more people need commodities and services of all kinds and society must be organized to supply them: more energy, transportation, goods, money, waste disposal, shops, factories, and health care. All have their impacts and link population concerns to the economy, social issues, and politics.

For a country or a continent, how many people are enough and how many are too many? A difficult question! Some experts say a sustainable world population is less than half the current number; other experts claim technology will meet any need. Thinking of the overall needs is difficult, whether a state, a country, or the world. Easier to grasp is what a shepherd knows.

If the owner of a flock of goats has a pasture, he knows that the plot can support only so many animals. That number is called the *carrying capacity*. If that limit is exceeded, the land suffers from overgrazing. If the owner of the flock needs more animals to support a growing family, could he find and afford more pasture?

Likewise, from a local to a global outlook, the availability of land, water, and air will set limits to growth. Any region, small or large, has a carrying capacity. However, unlike plants for food for the goats in the pasture, wise management can increase the capacity.

Land can be a limiting factor. Land cannot be created, but land can be ruined, overgrazed, or submerged in reservoirs behind newly built dams. Poorly managed land in dry areas can turn to desert. By contrast, and unlike goats overgrazing a pasture, combining fertilizer and irrigation with better crops makes the land more productive.

Water can also be a limiting factor. Water cannot be created but can be ruined by pollution. When the monsoon falls short, crops fail. Water from a well is dependable unless the well runs dry. Irrigation is more dependable, but dams, canals, and pumps are expensive to install and maintain.

Even adequate water needs regulation, and rivers that cross political boundaries require special agreements. An example is River Cauvery, which originates in Karnataka and flows through Tamil Nadu on its way to the Bay of Bengal. The two states have a long-standing disagreement over the water sharing from the river, mainly about amounts used for irrigation. Karnataka, in 1928, had only 7% of the irrigated area in the two states. By 2000, the total area had tripled, but Karnataka's portion had jumped to 45%. The two states have been quarrelling at all levels since the early 1970s and before about their water allocation. Some of the protests have been very serious. As per Cauvery Tribunal dated 19 February 2013, Karnataka will get 37%, Tamil Nadu 58%, Kerala 4%, and Puducherry 1%.

The desire for water drives conflicts on other rivers within India and notably also those that leave India—the Indus to Pakistan and the River Ganga to Bangladesh. In his book *India After Gandhi*, Ramachandra Guha says: "Water, more than oil, is the resource most crucial for India's economic development—crucial both for agriculture and for sustaining the burgeoning population of the cities."

Like land and water, air can also be a limiting factor. When air quality drops, life suffers, sometimes dangerously. People in some parts of China, including the capital, Beijing, cope with extreme air pollution. What is the result? People wear masks, respiratory diseases increase, and life span drops. India has its air problems, some worse than China. Burning wood in the countryside contributes heavily to air pollution. In the cities and towns, older, inefficient vehicles stuck in congested traffic add to the haze. Delhi ranks the worst polluted city in terms of air quality. While the emission per person is low, the total is high. Good practices can improve the air. China now generates more carbon dioxide than any other country, overtaking the United States. After China and the United States, India emits the most carbon dioxide. Around the earth, atmospheric carbon dioxide now exceeds 400 parts per million (ppm), the highest level in 2.5 million years. As carbon dioxide is a greenhouse gas, a major cause of global climate change, the effect is worldwide.

Poor or insufficient land and water trouble rural areas, and other trends strain the urban environment. People are moving from rural areas to cities, jamming what is already crowded. Even with a stable population, as more people demand more energy and can afford vehicles, emissions will rise, carbon dioxide included. When the population increases, the effects also increase.

A look at Bangladesh's environmental difficulties can help India recognize its problems. Circumstances intertwine the two countries; the monsoon drenches both, and they share a history from independence and before. River Ganga empties into Bay of Bengal flowing south and east through the Gangetic Plain of North India into Bangladesh. Bangladesh attained its independence in 1971. From that time, its population has more than doubled, and because of a high birth rate, it continues to grow.

As Bangladesh is sandwiched between West Bengal and the north-eastern states of India, the two countries face connected environmental problems. Some should be manageable. Management can be cooperative across the watery border in the Sunderbans, the world's largest mangrove forest.

However, there is trouble in other relations. If India pollutes the river waters or stores them in a dam, Bangladesh suffers. As already mentioned, Bangladesh is the world's most densely populated country. When disaster strikes, refuge is hard to find, but India is close. Most of Bangladesh lies low, 12 m or less above sea level. Danger comes from big rivers flooding and from the drenching storm surges of powerful cyclones. Bangladesh will feel the impact of rising sea levels as much as any country. A 1 m rise caused by climate change could happen in this century, flooding 10% of Bangladesh. If that happens, there will be "environmental refugees", driven from the swamped lands. From a crowded country, will they try to resettle in India? There is precedent. In the 1960s, about 60,000 people were displaced by the construction of a dam in what was still East Pakistan. They relocated in Arunachal Pradesh, where they remain as second-class citizens.

In his book *Collapse*, Jared Diamond catalogues civilizations, large and small, that have fallen apart. Spread across the globe from the North Atlantic to the South Pacific, the failures had a common thread, the misuse of resources. The lesson is clear. We have only this planet earth; if we overuse our life support, whether land, water, or air, global civilization could collapse.

6

Pollution, Sources, and Effects

6.1 AIR POLLUTION

6.1.1 Introduction

Air is essential for our body as we get oxygen from air to breathe. Air is 99.9% nitrogen, oxygen, water vapour, and inert gases. Clean air is hardly found in urban areas because of natural and human-made pollution. Human activities release some substances into the air that cause problems for all living beings.

Any substance in the air that can harm living organisms and the "built environment" is regarded as an air pollutant, which may be natural or human made. It can be in the form of solid particles, liquid droplets, or gases. Pollutants can be classified as primary and secondary. Each pollutant has serious implications for our health and well-being, as well as for the whole environment.

6.1.2 Primary Pollutants

6.1.2.1 Particulate matter

Particulate matter is made up of tiny solid or liquid particles suspended in air. Sources of particulate matter can be human made or natural. Some particulates occur naturally, originating from volcanoes, dust storms, forest and grassland fires, and sea spray. Human activities such as burning of fossil fuels in vehicles, power plants, and various industries also generate significant amounts of aerosols. Increased levels of fine particles in the air have been linked to health hazards such as heart disease, altered lung function, and lung cancer.

6.1.2.2 Carbon monoxide

Carbon monoxide is a colourless, odourless, and non-irritating gas, but it is very poisonous. It is produced by the incomplete combustion of fuels such as natural gas, coal, or wood. It is released into the atmosphere mainly from automobile exhaust.

6.1.2.3 Sulphur dioxide

Sulphur dioxide is one of the principal contaminants of air. It originates from the combustion of sulphur contained in fossil fuels. It is present in huge quantities in areas where coal is used as a fuel, such as electric power plants. Sulphur dioxide is also released from smelters where the sulphur in an ore is roasted, such as in copper, lead, and zinc smelting industries. Oil refineries, sulphuric acid manufacturing industries, fertilizer industries, and paper and pulp industries give out significant amounts of sulphur dioxide. Smelting operations are reported to cause heavy damage to agricultural and forest areas. Since sulphur dioxide is absorbed by water surfaces, it is harmful to soil and vegetation and causes deterioration and corrosion of materials such as metals, paper, paints, and leather.

Further oxidation of SO_2, usually in the presence of catalyst NO_2, forms H_2SO_4 and thus acid rain. This is one of the major concerns for the environmental impact of the use of these fuels as power sources.

6.1.2.4 Nitrogen oxides

The oxides of nitrogen are the second most abundant atmospheric pollutants. The primary source of oxides of nitrogen is automobile exhaust. These are also generated as by-products in chemical industries producing nitric acid, sulphuric acid, and nylon intermediates. Nitrogen dioxide is emitted from high-temperature combustion and is also produced naturally during thunderstorms by electrical discharge. It is one of the prominent air pollutants. The oxides of nitrogen are extremely dangerous to human health. The effects are sometimes more severe than those of carbon monoxide.

6.1.2.5 Hydrogen fluoride

Hydrogen fluoride and other volatile fluorides are considered serious pollutants even when they are present in concentrations of 0.001 parts per million (ppm) by volume. Fluorides are liberated mainly from aluminium smelting industries. The manufacture of phosphate fertilizers and ceramics, as well as some foundry operations, also contributes to hydrogen fluoride. Silicon tetrafluoride and hydrogen fluoride are reported to be toxic to some plants in concentrations lower than 0.1 parts per billion (ppb). Hydrogen fluoride accumulates in the leaves of plants and thereby causes fluorosis in animals. Leaves and flowers of many plants are extremely susceptible to fluorides. Hence, it is a major problem for agriculture units located in the vicinity of aluminium processing industries. Moreover, fluorides have the capacity to etch glass and can, therefore, cause considerable destruction of the material.

6.1.2.6 Volatile organic compounds

Volatile organic compounds (VOCs) are an important outdoor air pollutant. They are often divided into separate categories of methane (CH_4) and non-methane VOCs. Methane is an extremely potent greenhouse gas (GHG) that contributes to enhanced global warming. Other hydrocarbon VOCs are also significant GHGs via their role in creating ozone and prolonging the life of methane in the atmosphere, although the effect varies depending on local air quality. Within the non-methane VOCs, the aromatic compounds benzene, toluene, and xylene are suspected carcinogens and may lead to leukaemia through prolonged exposure. 1,3-butadiene is another hazardous compound often associated with industrial uses. These VOCs react with primary anthropogenic pollutants, specifically oxides of nitrogen, sulphur dioxide, and anthropogenic organic carbon compounds, to produce a seasonal haze of secondary pollutants.

6.1.2.7 Chlorofluorocarbons

Chlorofluorocarbons (CFCs) are harmful to the ozone layer and are emitted from products currently banned from use. Since the international ban on the use of CFCs, the hole in the ozone layer has reduced in size.

6.1.2.8 Ammonia

Ammonia is a compound normally encountered as a gas with a characteristic pungent odour. It is widely used as a fertilizer. Ammonia, directly or indirectly, is also a building block for the synthesis of many pharmaceuticals. Although ammonia is widely used, it is both caustic and hazardous.

6.1.2.9 Aldehydes and organic acids

Aldehydes and organic acids are present in lower concentrations in the atmosphere. The incomplete combustion of petroleum fuels and the incomplete oxidation of lubricating oils result in the formation of these chemicals. Combustion of natural gas may also contribute to the formation of these materials.

6.1.3 Secondary Pollutants

Secondary pollutants are formed in the air by the interaction of primary pollutants among themselves or the reaction with normal atmospheric constituents, such as sunlight and water vapour, with or without photoactivation. Experimental evidence indicates that exhaust gases from automobiles have particular importance in the formation of secondary pollutants, such as smog and ground-level ozone.

6.1.3.1 Smog

Classic smog results from large amounts of coal burning in an area and is caused by a mixture of smoke and sulphur dioxide. Modern smog is not usually produced from coal burning, but it results from vehicular and industrial emissions. These emissions are acted on in the atmosphere by ultraviolet light from the sun to form secondary pollutants, which also combine with primary emissions to form photochemical smog.

6.1.3.2 Ground-level ozone

Ozone is formed when nitrogen oxides emitted by the combustion of petroleum products interact with sunlight. Note that ozone is not emitted to the atmosphere but is formed only when primary pollutants interact among themselves and other elements in the atmosphere. It is a key constituent of the higher atmosphere that filters harmful ultraviolet radiation. However, ozone is a secondary pollutant at the ground level.

$$NO_2 \longrightarrow NO + O$$

Nitrogen dioxide   Nitric oxide   Atomic oxygen

$$O_2 + O \longrightarrow O_3$$

Molecular oxygen   Atomic oxygen   Ozone

$$NO + O_2 \longrightarrow NO_3$$

Nitrogen oxygen   Molecular oxygen   Nitrogen trioxide

$$NO_3 + O_2 \longrightarrow NO_2 + O_3$$

Nitrogen trioxide   Molecular oxygen   Nitrogen dioxide   Ozone

$$2NO_4 + O_3 \longrightarrow N_2O_5 + O_2$$

Nitrogen dioxide   Ozone   Nitrogen pentoxide   Molecular oxygen

6.1.4 Sources of Air Pollution

Sources of air pollution can be classified into two major categories: natural and anthropogenic or human made.

6.1.4.1 Natural sources

Some important natural sources of air pollution are as follows:

- Dust from large areas of land with little or no vegetation
- Poisonous gases such as sulphur dioxide, hydrogen sulphide, and carbon monoxide released in the atmosphere by volcanic eruption
- Methane emitted by the digestion of food by animals such as cattle
- Smoke and carbon monoxide from wild forest fires
- Carbon dioxide and methane produced from seed germination, marsh gas production, decomposition, and biodegradation of organic matter
- Volatile organic compounds emitted by vegetation in some areas on warmer days

6.1.4.2 Human-made sources

Some human-made sources of air pollution include deforestation, combustion of fossil fuels, transportation, agricultural activities, and industrialization.

Deforestation: Plants maintain the balance between carbon dioxide and oxygen in nature. Forests have a specific place in nature. Indiscriminate cutting of plants and trees in forests is termed deforestation. The main reasons for deforestation are rapid explosion of human and livestock population, expansion of agricultural croplands for catering to the increased population, enhanced requirement of timber, and growth of industries. The serious consequences of deforestation are as follows:

- Owing to deforestation, the concentration of carbon dioxide in the atmosphere increases, which produces greenhouse effect.
- Deforestation leads to the decrease in the availability of timber for the construction of buildings.
- Agricultural productivity decreases because of soil erosion.
- Frequent flooding erodes earth's surface.
- Oxygen depletion in the atmosphere occurs.

To avoid these negative consequences, forest lands should be protected and harvested lands should be reforested.

Combustion of fossil fuels: Coal, natural gas, and petroleum are organic materials and are called fossil fuels. They generate energy on combustion and cause air pollution. Burning of fossil fuels generates smoke, which contains poisonous gases such as carbon monoxide, sulphur dioxide, and unburnt black carbon particles. Coal and oil also contain sulphur as impurity and release SO_2 and CO_2 when they are burnt.

Transportation: Modern transportation consists of cars, trains, taxis, trucks, scooters, buses, lorries, and airplanes. CO, NO, and NO_2 released by internal combustion in engines cause air pollution. Automobile exhaust is responsible for more than 75% of the total air pollution.

Agricultural activities: Pesticides and fertilizers used in agriculture cause air pollution, especially when these are sprayed.

Industrialization: Chemical industries, paper mills, cotton mills, metal extraction plants, petroleum refineries, and other industries emit various poisonous pollutants (as gases) into the air. The Bhopal gas tragedy, which took place on 3 December 1984 in India, is considered the worst chemical disaster in history. Methyl isocyanate vapours discharged into the atmosphere by an explosion in the Union Carbide's pesticide factory killed thousands and injured many people.

Table 6.1 gives the sources of important air pollutants and their effects.

Table 6.1 Sources of important air pollutants and their effects on living beings

Pollutants	Major sources	Typical effects
Carbon monoxide (CO)	Incomplete combustion of fuels, automobile exhaust, jet engine emissions, blast furnaces, mines, and tobacco smoking	Toxicity, blood poisoning, increased proneness to accidents, and central nervous system impairment
Sulphur dioxide (SO_2)	Combustion of coal and petroleum products, burning of refuse, petroleum industry, oil refining, power houses, sulphuric acid plants, metallurgical operations, and domestic burning of fuels	Increased breathing rate and feeling of air starvation, suffocation, aggravation of asthma and chronic bronchitis, impairment of pulmonary functions, respiratory irritation, sensory irritation, irritation of throat and eyes
Oxides of nitrogen (NO_x)	Automobile exhausts, coal-fired and gas-fired furnaces, boilers, power stations, explosive industry, fertilizer industry, manufacture of HNO_3, combustion of wood and refuse	Respiratory irritation, headache, bronchitis, pulmonary emphysema, impairment of lungs, lachrymatory effect, loss of appetite, corrosion of teeth
Hydrogen sulphide (H_2S)	Coke ovens, kraft paper mills, petroleum industry, oil refining, viscose, rayon manufacturing plants, manufacture of dyes, tanning industry, and sewage treatment plants	Headaches, conjunctivitis, sleeplessness, pain in the eyes, irritation of respiratory tract, respiratory paralysis, asphyxiation, malodorous; blockage of oxygen transfer, poisoning cell enzymes, and damaging nerve tissues in the case of high concentrations

Contd...

Table 6.1 *contd...*

Pollutants	Major sources	Typical effects
Chlorine (Cl_2)	Accidental breakage of chlorine cylinders, electrolysis of brine, bleaching of cotton pulp, and other process industries using chlorine	Irritation to eyes, nose, and throat, toxicity, respiratory irritation, lachrymatory effects; oedema, pneumonitis, emphysema, and bronchitis in large doses
Hydrogen fluoride (HF)	Glass fibre manufacture, chemical industry, fertilizer industry, aluminium industry, ceramic industry, phosphate rock processing	Irritation, respiratory diseases, fluorosis of bones, mottling of teeth
Carbon dioxide (CO_2)	Combustion of fuels, automobile exhausts, jet engine emissions	Toxic in large quantities, hypoxia
Hydrocarbons (HC)	Organic chemical industries, petroleum refineries, automobile exhausts, rubber manufacture	Some hydrocarbons have carcinogenic effects, lachrymatory effect
Oxidants (such as O_3)	Photochemical reactions in atmosphere involving organic materials and NO_2, reactions induced by silent electrical discharge and intense ultraviolet radiations in the atmosphere	Irritation of lungs, eyes, and respiratory tract; accumulation of fluids in lungs and damage to lung capillaries. These biochemical effects of O_3 mostly arise from generation of free radicals, which attack the SH groups present in enzymes
Dust	Mining activities, asbestos factories, power stations, metallurgical industries, ceramic industry, factory stacks, glass industry, cement industry, foundries	Respiratory diseases, toxicity from metallic dust, silicosis and asbestosis from specific dusts. Asbestos dust causes pulmonary fibrosis, pleural calcification, and lung cancer
Ammonia (NH_3)	Chemical industries, coke oven refineries, stockyards, fuel incineration	Damage to respiratory tracts and eyes, corrosive to mucous membranes
Formaldehyde (HCHO)	Waste incineration, automobile exhausts, combustion of fuels, photochemical reactions	Irritation to eyes, skin, and respiratory tracts
Arsenic (As)	Arsenic containing fungicides, pesticides, and herbicides, metal smelters, by-product of mining activities, chemical wastes	Inhalation, ingestion, or absorption through skin can cause mild bronchitis, nasal irritation, or dermatitis. Carcinogenic activity is also suspected. Attack SH groups of enzymes, coagulate proteins

Contd...

Table 6.1 *contd...*

Pollutants	Major sources	Typical effects
Cadmium (Cd)	Cadmium-producing industries, electroplating, welding; by-products from refining of Pb, Zn, and Cu, fertilizer industry, pesticide manufacture, cadmium–nickel batteries, nuclear fission plants, production of TEL (tetraethyl lead) used as the additive in petrol	Inhalation of fumes and vapours causes kidney damage, bronchitis, gastric and intestinal disorders, cancer, disorder of heart, liver, and brain, chronic and acute poisoning. Renal dysfunction, anaemia, hypertension, bone marrow disorder, and cancer
Chromium (Cr)	Metallurgical and chemical industries, processes using chromate compounds, cement and asbestos units	Toxic to body tissues, cause irritation, dermatitis, ulceration of skin, perforation of nasal septum. Carcinogenic action suspected
Lead (Pb)	Automobile emission, lead smelters, burning of coal or oil, lead arsenate pesticides, smoking, mining, and plumbing	Absorption through gastrointestinal and respiratory tract and deposition in mucous membranes, cause liver and kidney damage, gastrointestinal damage, mental retardation in children, abnormalities in fertility and pregnancy
Zinc (Zn)	Zinc refineries, galvanizing processes, brass manufacture, metal plating, plumbing	Zinc fumes have corrosive effects on skin and can cause irritation and damage mucous membranes
Manganese (Mn)	Ferromanganese production, organomanganese fuel additives, welding rods, incineration of manganese-containing substances	Poisoning of the central nervous system, absorption, ingestion, inhalation or skin contact may cause manganic pneumonia.
Nickel (Ni)	Metallurgical industries using nickel, combustion of fuels containing nickel additives, burning of coal and oil, electroplating using nickel salts, incineration of nickel-containing substances, vanaspati manufacture	Respiratory disorders, dermatitis, cancer of lungs and sinus
Mercury (Hg)	Mining and refining of mercury, organic mercurials used in pesticides, laboratories using mercury	Inhalation of mercury vapours may cause toxic effects and protoplasmic poisoning. Organic mercurials are highly toxic and may cause irreversible damage to nervous system and brain

6.1.5 Indoor Air Pollution

Pollution also needs to be considered inside our houses, offices, and schools. Some pollutants can be created by indoor activities such as smoking and cooking. We spend maximum of our time inside buildings, and so our exposure to harmful indoor pollutants can be serious. It is, therefore, important to consider indoor pollution. Several household products such as cleaners, detergents, paints, disinfectants, and insecticides pose a serious threat to indoor air quality.

6.2 WATER POLLUTION

6.2.1 Introduction

Water is a unique substance found in nature since it cannot be manufactured. The pressure of increasing population, growth of industries, urbanization, energy-intensive lifestyle, loss of forest cover, and lack of environmental awareness are causing significant water pollution. The lack of implementation of environmental rules and regulations and environment improvement plans, untreated effluent discharge from industries and municipalities, use of non-biodegradable pesticides/fungicides/herbicides/insecticides, and the use of chemical fertilizers instead of organic manures are some other causes of water pollution. Humans are using more and more materials that pollute the drinking water sources. They are dumping contaminants into the small portion of water on the planet that is fit for drinking.

The earth's surface is 75% water, but only 3% is freshwater, of which only 1% is available for human use (2% is locked up in icebergs). Water sustains life for humans, animals, and plants. People need water for basic everyday activities, such as drinking and cooking, but water is also important for sustaining agriculture and industry. However, the supply of freshwater available to humanity is shrinking because of the pollution of freshwater resources. Lakes and rivers have become polluted with an assortment of waste, including untreated or partially treated municipal sewage, toxic industrial effluents, harmful chemicals, and run-off from agricultural activities. Polluted water supplies not only limit water availability but also put millions at risk of water-related diseases.

In 1995, the Central Pollution Control Board (CPCB) identified severely polluted stretches on 18 major rivers in India. Not surprisingly, a majority of these stretches were found in and around large urban areas. The high incidence of severe contamination near urban areas indicates that the contribution of industrial and domestic sectors to water pollution is much higher than their relative importance in the Indian economy.

Agricultural activities also contribute in terms of overall impact on water quality. Besides a rapidly depleting groundwater table in different parts, the country faces another major problem: groundwater contamination. This problem has affected as many as 19 states, including Delhi. Geogenic contaminants, including salinity, iron, fluoride, and arsenic, have affected groundwater in more than 200 districts spread across 19 states.

As an environmental resource, water is regenerative: that is, it can absorb pollution load up to certain levels without affecting its quality. In fact, water pollution occurs only when pollution loads exceed the natural regenerative capacity of a water source. The control of water pollution, therefore, entails reduction in the pollution loads from anthropogenic activities to the natural regenerative capacity of the source. The benefits of preserving water quality are manifold. The abatement of water pollution not only provides marketable benefits such as reduced waterborne diseases, savings in the cost of supplying water for household, industrial, and agricultural uses, control of land degradation, and development of fisheries, but also generates non-marketable benefits such as improved environmental amenities, aquatic life, and biodiversity.

Water pollution is a major global problem that requires ongoing evaluation and revision of water resource policy at all levels (from international to individual aquifers and wells). It has been suggested that it is the leading worldwide cause of deaths and diseases and that it accounts for the deaths of more than 14,000 people daily. In addition to the acute problems of water pollution in developing countries, industrialized countries continue to struggle with pollution problems as well. Water is typically referred to as polluted when it is impaired by anthropogenic contaminants and either does not support human use, such as drinking, or undergoes a marked shift in its ability to support its constituent biotic communities, such as fish. Natural phenomena such as volcanoes, algae blooms, storms, and earthquakes also cause major changes in water quality and the ecological status of water.

Most water pollutants are eventually carried by rivers into the oceans. In some regions of the world, studies using hydrology transport models traced the influence of pollutants up to 100 miles from the river mouth. The specific contaminants leading to pollution in water include a wide spectrum of chemicals, pathogens, and physical or sensory changes, such as elevated temperature and discoloration. Although many of the chemicals and substances that are regulated may be naturally occurring (calcium, sodium, iron, manganese), the concentration is often the key in determining what is a natural component of water and what is a contaminant. High concentrations of naturally occurring substances can have negative impacts on aquatic flora and fauna. Oxygen-depleting substances may be natural materials, such as plant matter (leaves and grass), or man-made chemicals. Other natural and anthropogenic substances may cause turbidity (cloudiness), which blocks light and disrupts plant growth, clogging the gills of some fish species.

Many of the chemical substances are toxic. Pathogens can produce waterborne diseases in human or animal hosts. Alteration of water's physical chemistry includes acidity (change in pH), electrical conductivity, temperature, and eutrophication. Eutrophication is the increase in the concentration of chemical nutrients in an ecosystem to the extent that increases the primary productivity of the ecosystem. Depending on the degree of eutrophication, subsequent negative environmental effects such as anoxia (oxygen depletion) and severe reduction in water quality may occur, affecting fish and other animal populations.

An estimated 700 million Indians have no access to proper toilets, and around 1000 Indian children die of diarrheal illness every day. Some 90% of China's cities suffer from some degree of water pollution, and nearly 500 million people lack access to safe drinking water. In addition to the acute problem of water pollution in developing countries, developed countries continue to struggle with pollution problems as well. In the most recent national report on water quality in the United States, 45% of assessed stream miles, 47% of assessed lake acres, and 32% of assessed bays and estuarine square miles have been classified as polluted.

When toxic substances enter lakes, streams, rivers, oceans, and other waterbodies, they get dissolved or lie suspended in water or get deposited on the bed. Pollutants can also seep down and affect the groundwater deposits. Water pollution has many sources. The most polluting of them are the city sewage and industrial waste discharged into rivers. Facilities to treat wastewater are not adequate in any Indian city. At present, only about 10% of the wastewater generated is treated, and the rest is discharged into waterbodies. Owing to this, pollutants enter groundwater, rivers, and other waterbodies. Such water, which ultimately ends up in households, is often highly contaminated and carries disease-causing microbes. Agricultural run-off from fields, which drains into rivers, is another major water pollutant as it contains fertilizers and pesticides.

6.2.2 Types of Water Pollutants

6.2.2.1 *Industrial pollutants*

Wastewater from manufacturing or chemical processes in industries contributes to water pollution. Industrial wastewater usually contains specific and readily identifiable chemical compounds. During the last 50 years, the number of industries in India has grown rapidly. But water pollution is concentrated within small- and medium-scale industries that do not have adequate wastewater treatment facilities. Of this, a large portion can be traced to the processing of industrial chemicals and food processing industry. In fact, a number of small- and medium-sized industries in the region covered by the Ganga Action Plan do not have adequate effluent treatment facilities.

6.2.2.2 Biological pollutants

Pathogens: Coliform bacteria are a commonly used indicator of water pollution, although not an actual cause of disease. Other microorganisms sometimes found in surface water, which cause human health problems, include the following:

- *Burkholderia pseudomallei*
- *Cryptosporidium parvum*
- *Giardia lamblia*
- *Salmonella*
- *Norovirus* and other viruses
- Parasitic worms (helminths)

High levels of pathogens may result from inadequately treated sewage discharges. This can be caused by a sewage plant designed with minimal secondary treatment (more typical in less developed countries). In developed countries, older cities with ageing infrastructure may have leaky sewage collection systems (pipes, pumps, valves), which can cause sanitary sewer overflows. Some cities also have combined sewers, which may discharge untreated sewage during rainstorms. Figure 6.1 gives an overview of polluted water streams.

6.2.2.3 Chemical and other contaminants

Contaminants may include organic and inorganic substances. Organic water pollutants include the following:

Figure 6.1 An overview of polluted water streams

- Detergents
- Disinfection by-products found in chemically disinfected drinking water, such as chloroform
- Food processing waste, which can include oxygen-demanding substances, fats, and grease
- Insecticides and herbicides, a huge range of organohalides, and other chemical compounds
- Petroleum hydrocarbons, including fuels (gasoline, diesel fuel, jet fuels, and fuel oil), lubricants (motor oil), and fuel combustion by-products, from storm water run-off
- Tree and bush debris from logging operations
- Volatile organic compounds such as industrial solvents from improper storage
- Chlorinated solvents, which are dense non-aqueous phase liquids that may fall to the bottom of reservoirs, since they do not mix well with water and are denser. They include the following:
  - Polychlorinated biphenyl
  - Trichloroethylene
  - Perchlorate

Inorganic water pollutants include the following:
- Acidity caused by industrial discharges (especially sulphur dioxide from power plants)
- Ammonia from food processing waste
- Chemical waste as industrial by-products
- Fertilizers containing nutrients (nitrates and phosphates), which are found in storm water run-off from agriculture, as well as commercial and residential use
- Heavy metals from motor vehicles (via urban storm water run-off) and acid mine drainage
- Silt (sediment) in run-off from construction sites, logging, slash-and-burn practices, or land-clearing sites

6.2.2.4 *Thermal pollutants*

Thermal pollution is the rise or fall in the temperature of a natural body of water caused by human influence. Thermal pollution, unlike chemical pollution, results in a change in the physical properties of water. A common cause of thermal pollution is the use of water as a coolant by power plants and industrial manufacturers. Elevated water temperature decreases oxygen levels, which can kill fish and alter the food chain composition, reduce species biodiversity, and foster invasion by new thermophilic species. Thermal pollution can also be caused by the release of very cold water from the base of reservoirs into warmer rivers.

6.2.3 Sources of Water Pollution

6.2.3.1 Point sources

Point source water pollution refers to contaminants that enter a waterway from a single identifiable source, such as a pipe or ditch (Figure 6.2). Examples of sources in this category include discharges from a sewage treatment plant, a factory, or a city storm drain. The U.S. Clean Water Act defines point sources for regulatory enforcement purposes and issues National Pollutant Discharge Elimination System (NPDES) permits.

Figure 6.2 An overview of point source pollution

6.2.3.2 Non-point sources

Non-point source (NPS) pollution refers to diffuse contamination that does not originate from a single discrete source. NPS pollution is often the cumulative effect of small amounts of contaminants gathered from a large area. A common example is the leaching out of nitrogen compounds from fertilized agricultural lands. Nutrient run-off in storm water from "sheet flow" over an agricultural field or a forest is also cited as an example of NPS pollution.

Contaminated storm water from parking lots, roads, and highways, called urban run-off, is sometimes included under the category of NPS pollution. However, this run-off is typically channelled into storm drain systems and discharged through pipes to local surface waters and is a point source.

6.2.4 Water Pollution Quantification

6.2.4.1 Sampling

Sampling of water for physical or chemical testing can be done by several methods, depending on the accuracy needed and the characteristics of the contaminant. Many contamination events are sharply restricted in time, most commonly in association with rain events. For this reason, "grab" samples are often inadequate for fully quantifying contaminant levels. Scientists gathering this type of data often employ auto-sampler devices that pump increments of water at either time or discharge intervals. Sampling for biological testing involves collection of plants and animals from the surface waterbody. Depending on the type of assessment, the organisms may be identified for biosurveys (population counts) and returned to the waterbody, or they may be dissected for bioassays to determine toxicity.

6.2.4.2 Physical testing

Common physical tests of water include temperature, solids concentrations (total suspended solids), and turbidity.

6.2.4.3 Chemical testing

Water samples may be examined using the principles of analytical chemistry. Many published test methods are available for both organic and inorganic compounds. Frequently used methods include pH, biochemical oxygen demand, chemical oxygen demand, nutrients (nitrate and phosphorus compounds), metals (copper, zinc, cadmium, lead, and mercury), oil and grease, total petroleum hydrocarbons, and pesticides.

6.2.4.4 Biological testing

Biological testing involves the use of plant, animal, and microbial indicators to monitor the health of an aquatic ecosystem. Several tests are available, and these are getting more accurate with increases in measurement technologies.

6.2.5 Water Pollution and Health Aspects

6.2.5.1 Water pollution effects on animals

While humans feel the harmful consequences of water pollution only when they consume contaminated water, go swimming in polluted water sites, or catch aquatic life forms from polluted waterbodies, animals are easier victims of the harmful effects of water pollution. While humans can treat polluted water to make it safe and drinkable and can always choose not to bathe in polluted waters or refrain from eating aquatic animals, animals do not have any of these alternatives to escape the toxicity of water that has been contaminated by human and industrial waste. Some common issues faced by animals from contamination of water by humans are as follows:

- Chemical contaminants carried by industrial wastes kill a lot of smaller aquatic organisms such as frogs, fish, and tadpoles. This, in turn, causes a loss of food sources for bigger aquatic creatures, leading them to either consume poisoned, dead fish and perish, or leave their natural habitat to go in search of food in other aquatic quarters. Often, this leads to sickness and death of these animals.

- An excess of nutrients such as nitrogen and phosphorus in the water leads to an increased growth of toxic algae and aquatic plants that cause poisoning and death in fish and other animals.

- The presence of huge quantities of mercury in water has led to a lot of undesirable changes in aquatic species. Too much mercury leads to hormonal imbalances and glandular damage, leading to abnormal behavioural shifts. Also, mercury is a toxic metallic chemical that impacts the reproductive functions, growth, and development of animals continuously exposed to high doses.

- Oil spills that introduce unhealthy amounts of oil into the marine environment also make marine animals sick and lead to their unnatural death. Dumping solid trash such as plastic, metallic scrap, and garbage may block aquatic channels and can also cause small animals to get trapped in the debris. Most water-dwelling animals tend to suffocate or drown on being trapped and are unable to swim.

- Polluted water used for irrigation also contaminates the soil and the agricultural produce. This may lead to health issues in herbivorous animals that feed on agricultural plants and leftovers. These pollutants can radically alter the metabolism of a number of soil-dwelling bacteria and insects, making them perish or unsuitable for consumption by common predators of the local ecosystem. Atmospheric pollutants may get mixed with clouds and fall back on earth as acid rain. This toxic shower is potent enough to inflict mortal injuries to any life form that gets exposed to it.

These are a few of the most common and prominent repercussions of water pollution on animals. These consequences demonstrate that most animals are more affected than humans, when water is contaminated. Although various measures need to be taken on national and industrial levels to arrest this situation before it goes totally out of control, individuals can do their small part by refraining from littering beaches and lakes with paper, plastic, and other garbage so that lesser animals make it to the extinct species list, and people can tell their children that the ocean waters still abound in whales, dolphins, turtles, seals, and all those other animals that are marked as endangered species.

6.2.5.2 Water pollution effects on humans

Water acts as a purifier in living bodies. If enough water is not consumed, the body cannot properly flush out unhealthy toxins from the kidneys and liver and bowels are not properly and completely emptied. The longer the waste remains in the body, the more time the body gets to reabsorb the toxins back into the bloodstream. As a result, the toxins make way through the human body, causing poisoning and spreading infections. According to medical experts, an individual needs to consume at least 2 L of water daily for survival. The health and livelihood of populations worldwide depend on the availability of a safe drinking water supply. As the human population increases and modern technology develops further, the risk of water contamination also increases. Examples of water contamination sources include animal and human wastes, chemicals disposed of improperly, and landfills. If not treated properly, drinking water can pose a severe health risk for humans (Figure 6.3).

Waterborne diseases: Waterborne diseases "arise from the contamination of water by human and/or animal body excretions infected by pathogenic viruses or bacteria, which are directly transmitted when the water is consumed or used for food preparations". Examples of waterborne diseases include cholera, typhoid, and cryptosporidiosis.

Water-privation diseases: Water-privation diseases are "affected by the quantity of water". The disease is spread through (infected) person-to-person contact or through contact with infected materials. Poor personal hygiene is a common factor leading to water-privation diseases.

Water-based diseases: In water-based diseases, "water provides the habitat for intermediate host organisms in which parasites are able to spend part of their life cycle and later their infective larval forms in water are passed on to humans".

Water-related diseases: In water-related diseases, water provides a "home for insects". Examples of these types of diseases include malaria, dengue, and yellow fever.

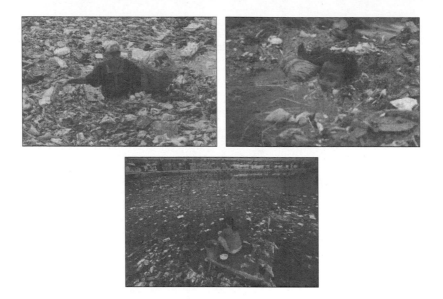

Figure 6.3 Impact of water pollution on human beings

Water-dispersed infections: In these types of infections, "pathogens are able to proliferate in freshwater and enter the body through the respiratory tract".

6.2.6 Pollution Control Measures

In order to avoid the ill effects of water pollution on human and animal health and agriculture, standards/rules/guidelines have been devised for the discharge of effluents from industries and municipalities, quality of drinking water, irrigation water, criteria for aquatic life in freshwater by various authorities, including the Indian CPCB, World Health Organization, World Bank, Indian Standard Institution, and Indian Council of Medical Research. The implementation of these rules, standards, and guidelines is, however, inadequate. Improperly treated or even untreated industrial and municipal effluents have been continuing to pollute not only surface water sources but also groundwater. Alarming levels of lead (Pb) and arsenic (As) have been found in the sediments and waters of Damodar, Safi, Ganga, and Adjai rivers in Jharkhand and West Bengal.

6.2.7 Techniques for Water Treatment

6.2.7.1 Primary treatment

In today's modern treatment plants, first wastewater receives primary treatment. During the primary treatment, solids found in raw wastewater are either screened out or allowed to settle at the bottom of the tank. Solids removed from the bottom of the tank are called primary sludge.

6.2.7.2 Secondary treatment

Secondary treatment started on a large scale in the United States in the early 1970s. In the secondary treatment, wastewater flows from primary treatment tanks to larger secondary treatment tanks, during which a number of processes occur:

- Large amounts of bacteria and other microorganisms, similar to those found in streams, are mixed with the wastewater. Microorganisms use pollutants in wastewater as food and multiply very quickly.
- After the pollutants in wastewater are converted to microscopic organisms (through eating), the wastewater is kept for several hours in settling tanks.
- Organisms settle to the bottom of the tank, and clean water flows from the top. It is then disinfected and released to the receiving body of water (stream, river, lake, or ocean).

6.2.7.3 Biological treatment

The wastewater treatment process in most wastewater treatment plants depends on whether the pollutants in the sewage are biodegradable. A pollutant is biodegradable if there is a naturally occurring organism that can use the pollutant as food.

6.2.7.4 Phytoremediation

Plants are used to remove toxic substances during wastewater treatment, and the process is called phytoremediation.

6.2.7.5 Bioremediation

During wastewater treatment, microbes are utilized to remove or transform toxic substances through a process called bioremediation.

6.2.7.6 Other techniques

Flocculation: The process of flocculation involves removing dirt and other particles that are suspended in water. This is achieved by adding aluminium and iron salts to the water; these salts form sticky particles, which attract the particles to be removed.

Sedimentation: Once the particles are removed by flocculation, they naturally settle out of the water.

Filtration: Filtration is used to remove particles such as clays, organic matter, and chemicals and precipitates out of the water. The process of filtration purifies drinking water, thus decreasing possibility of contamination.

Disinfection: The process of disinfection is one of the most popular and advanced treatment methods of the 20th century. Disinfection is carried out before water has a chance to enter the distribution centre to ensure that the water is free from toxins. Effective and efficient sources of disinfectants include chlorine, chlorinates, and chlorine dioxides.

6.2.8 Conclusion

Water pollution is a serious problem in India as almost 70% of its surface water resources and a growing percentage of its groundwater reserves are contaminated by biological, toxic, organic, and inorganic pollutants. In many cases, these sources have been rendered unsafe for human consumption as well as for other activities such as irrigation and industrial needs. This shows that degraded water quality can contribute to water scarcity since it limits water availability for both humans and the ecosystem. Polluted water may have undesirable colour, odour, taste, turbidity, organic matter, harmful chemical content, toxic and heavy metals, pesticides, oily matters, industrial waste products, radioactivity, high total dissolved solids, acids, alkalis, domestic sewage, virus, bacteria, protozoa, rotifers, and worms. The organic content may be biodegradable or non-biodegradable. Pollution of surface water (rivers, lakes, and ponds), groundwater, and sea water is harmful for human and animal health. Pollution of drinking water and the food chain is, by far, the most worrisome aspect.

6.3 NOISE POLLUTION

Noise pollution is regarded as environmental noise or unwanted sounds that are annoying and distracting. It is also known as sound pollution. There are about 25,000 hair cells in the human ear which create wave in one ear, corresponding to the sound in the environment, as a response to different levels of frequencies. With increasing intensity/pitch/loudness of sound, the cells get destroyed, decreasing our ability to hear high-frequency sounds.

Decibel (dB) is used as a measure of sound intensity level or sound pressure level. It is named after Alexander Graham Bell, the inventor of the telephone.

$$\text{Intensity level (dB)} = 10\log_{10} + \frac{\text{Intensity measured}}{\text{Reference intensity}}$$

$$dB = 10\log_{10}\left[\frac{I_m}{I_0}\right]$$

Apart from loudness, the frequency or pitch of the sound determines whether the sound is harmful or not. A modified scale called decibel dBA-takes pitch into account. The average noise levels of some sources are summarized in Table 6.2.

For humans, the normal level of tolerance is 80 dBA. Sound level above this is considered noise pollution. Most electronic devices and motors emit sound above 80 dBA. Amplified rock music is 120 dBA. The Ministry of Environment, Forests and Climate Change has issued noise pollution regulations under the Environment Protection Act, 1986 and the limits are as follows:

Sector	Industrial	Commercial	Residential
Limit	75 dB	65 dB	55 dB

Further, public address systems are not permitted from 10 p.m. to 6 a.m. Rules also establish zones of silence around schools, courts, and hospitals within a radius of 100 m.

Table 6.2 Average noise levels of common sound sources

Source	Noise level (dBA)
Threshold of audibility/hearing	0
Conservation (quiet)	20–30
Conversation (face to face)	60
Classroom teaching	55–60
Home appliances	65–75
Road traffic (medium)	70–80
Heavy road traffic	80–90
Inside cinema hall	85–95
Horns of vehicles	90–105
Rail engine at 15 m	97–105
Loudspeakers	100–120
Threshold of pain	130
Jet engine at 25 m	140
Deepawali crackers	125–160
Bomb explosion	190

6.3.1 Sources of Noise

The major sources of noise are summarized as follows:
- **Transportation sources:** Railways, road traffic, and air traffic.
- **Industrial sources:** Noise is generated in almost all industrial activities, such as thermal power generation, mineral processing, product fabrication and assembly, steel plants, cement plants, and smelters.
- **Public address systems:** Use of loudspeakers during occasions such as marriages, functions, and festivals.
- **Agriculture machinery:** Use of tractors, pump sets for pumping water from tube wells, dug wells, farm machines for agriculture such as harvesters.
- **Defence equipment:** Shooting practices, wars, bomb explosion, and operation of track-mounted equipment such as bulldozers and tanks.
- **Household sources:** Mixer grinder, lawn mowers, food blenders, vacuum cleaner, water pumps, and so on.

6.3.2 Effects of Noise Pollution

Noise affects human health in the following ways:
- **Physical effects:** Damage to eardrum, temporary impairment of hearing, and permanent deafness
- **Physiological effects:** Muscular strain, headache, eye strain, decreased colour perception, nervous breakdown, and cardiac pain
- **Psychological effects:** Emotional disturbance, depression, fatigue, frustration, irritation, and reduced efficiency

6.3.3 Control of Noise Pollution

Noise pollution can be controlled by reducing noise at the source, interrupting the path of noise, and protecting the receiver.

6.3.3.1 Noise control at source

Eliminating noise at the source is the most effective method for preventing noise pollution. Some examples are as follows:
- Reduction in noise generated by the mechanical vibration of a machine by damping or isolating the vibration by applying a damping material (such as rubber) to vibrating components.
- Reduction in the impact force by optimizing the impact distance and covering either or both impact surfaces by rubber.

- Modification of manufacturing design such as enclosing engine parts within proper noise insulating materials.
- Other ways of controlling noise at the source include purchasing quieter models, isolating vibrations, removing unnecessary sources, erecting barrier, and installing absorptive treatment.

6.3.3.2 *Noise control at path*

When the source cannot be made quiet, noise can be controlled by modifying the path. Some examples are as follows:

- Attenuating noise by moving the noise source away from sensitive areas.
- Suppressing noise from automobiles using silencers.
- Reducing noise around residential areas by planting trees in the form of a green belt.
- Reducing the transmission of noise using acoustic screens and barriers.
- Enclosing noisy machines in isolated buildings.

6.3.3.3 *Noise control at receiver*

If source control and path control do not prove effective, possibility of control at the receiver should be explored. Some examples are as follows:

- Using hearing protection devices such as ear plugs and ear muffs. These devices reduce the level of noise (by 10–55 dB) entering the outer and middle ears before it reaches the inner ear.
- Enclosing and relocating the receiver.

6.4 SOIL POLLUTION

6.4.1 Causes of Soil or Land Pollution

Soil pollution is caused by direct and indirect sources. Direct sources harm the soil much more than indirect sources. Examples of direct causes are poor management of solid and liquid domestic, industrial, and agricultural wastes, waterlogging, soil erosion, soil salination, and improper disposal of medical waste. Examples of indirect causes are acid rain and illegal disposal of radioactive substances. The main reasons for soil pollution are as follows:

(i) **Agriculture:** Waterlogging may occur when the drainage system of agricultural fields is not maintained properly. Waterlogging closes the passage of air to the soil, stops the growth of soil organisms, and makes the soil barren.

(ii) **Deforestation/shifting cultivation:** If forests are cleared by burning or cutting down trees/vegetation for cultivation of crops, it causes severe soil erosion. More damage occurs when the land involved is hilly.

(iii) **Injudicious use of chemical fertilizers:** Use of inorganic fertilizers increases nutrient contamination. Soil microbes reduce the nitrogen to nitrite ions, which enter the animal body through food or water. These ions, which are directly absorbed in the bloodstream, oxidize oxyhaemoglobin (the O_2 carrier) to methemoglobin. Since methemoglobin cannot carry oxygen, the animal ultimately dies.

(iv) **Pesticides:** Pesticides are chemicals used by farmers to protect their crops. Large amounts of pesticides in the soil interfere with the soil's metabolic process. Pesticides kill many non-targeted beneficial soil organisms such as earthworms. Thus, the soil becomes infertile.

Organochlorides (such as dichlorodiphenyltrichloroethane or DDT) are second-generation pesticides. They are non-biodegradable substances. They accumulate and magnify in the food chain and interfere with the calcium metabolism of birds. As a consequence, birds lay fragile, thin-shelled eggs. Pesticides accumulate in the fatty tissues of prey in high concentration. Predators that eat these prey also get killed. Thus, pesticides lead to poisoning of the ecosystem.

(v) **Opencast mining:** In opencast mining, trees in reserved forests are mercilessly cut down to open up the ore deposit. Loss of greenery results in land degradation, drought, and desertification if proper land reclamation is not completed. This is illustrated in Figure 6.4.

(vi) **Solid wastes from homes and industries:** Chemical, petroleum, and metal-related industries, dry cleaners, and gas stations produce hazardous wastes such as oils, battery metals, and organic solvents. These hazardous wastes contaminate soil and water resources.

(vii) **Acid rain:** It converts neutral soil to an acidic one over time.

6.4.2 Effects of Soil Pollution

The harmful effects of soil pollution are briefly described as follows:

- Fluorosis occurs as a result of consuming fluoride-containing maize and jowar crops. The fluoride is absorbed by crops from the fluoride-contaminated soil.

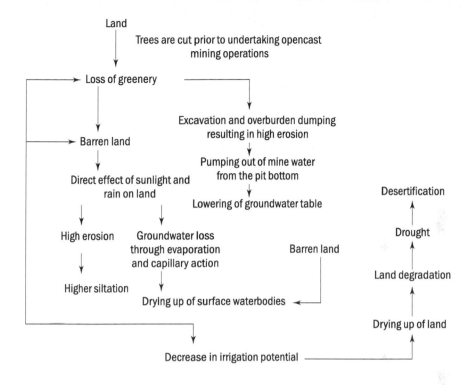

Figure 6.4 Land degradation cycle due to opencast mining, mineral dressing, and other related activities

- Toxic gases emitted from dumped municipal or other solid wastes on land are detrimental to health. The unpleasant smell and spread of insects cause inconvenience to people.
- Poisoning of the ecosystem.
- Contamination of underground and surface drinking water.
- Reduction in the fertility of soil.

6.4.3 Control and Treatment of Soil Pollution

Soils can be treated using several methods to remove harmful constituents; however, controlling pollution is more cost-effective. Some methods are as follows:

- Polluted soil can be treated using bioremediation. It uses microorganisms (yeast, fungi, or bacteria) to break down or degrade hazardous substances into less toxic or non-toxic substances (such as CO_2 and H_2O).

- The principle of three R's—recycle, reuse, and reduce—helps in minimizing the generation of solid and liquid wastes. In addition, using biofertilizers and natural pesticides helps in minimizing the usage of chemical fertilizers and pesticides, which harm soil in the long run.
- Proper disposal techniques must be employed. For example, composting of biodegradable solids and incineration of non-biodegradable solids should be done to prevent soil disposal.
- Planned afforestation helps prevent soil erosion.
- Formulation and effective implementation of stringent pollution control legislation can help in controlling soil pollution.
- Solid and liquid hazardous wastes and sludge must be treated before disposal on land.

6.5 SOLID WASTE POLLUTION

Solid waste is created every day in India, and much of this (up to 60%) is organic waste. Improper management of this waste impacts human health by polluting waterbodies and spreading diseases through flies and other insects. In addition, solid wastes clog city drainage systems by blocking drains. The Municipal Solid Waste (Management and Handling) Rules, 2000 mandate the segregation of waste at the source and the conversion of organic waste into useful compost or biogas. Many municipalities have now contracted their solid waste management to private companies that convert the waste to valuable products and minimize landfilling.

Many technologies are available for processing and converting municipal solid waste (MSW) into useful products for the society. These technologies convert waste into wealth. This chapter describes the status of these technologies in India. The technological options available for processing MSW are based on bioconversion or thermal conversion. The bioconversion process is applicable to the organic fraction of wastes to form compost or to generate biogas such as methane (energy) and residual sludge (manure). Various technologies are available, which can be broadly categorized into the following:

- Conventional and biological conversion methods include open dumping, sanitary landfilling/landfill gas recovery, composting, biomethanation/bio-waste derived fuel.
- Advanced thermal treatment technologies include incineration, combustion, pyrolysis and gasification, plasma pyrolysis, pelletization/refuse-derived fuel (RDF).

Each of these technologies has advantages and limitations. The technologies are briefly described in the following paragraphs, with a summary of the advantages and disadvantages.

Incineration is defined as the process of controlled combustion using an enclosed device to thermally break down combustible solid waste to an ash residue that contains little or no combustible material and that produces electricity, steam, or other energy as a result. Even though both incineration and RDF produce energy, the objective of incineration of MSW is to reduce its volume. Generating energy and electricity only adds value to this process. In other words, it is the process of direct burning of wastes in the presence of excess air (oxygen) at the temperature of about 1000°C and above, which liberates heat energy, inert gases, and ash. The net energy yield depends on the density and composition of the waste; relative percentage of moisture and inert materials, which add to the heat loss; ignition temperature; size and shape of constituents; and design of the combustion system (fixed bed/fluidized bed). In practice, about 65%–80% of the energy content of the organic matter can be recovered as heat energy, which can be utilized either for direct thermal applications or for producing power via steam turbine generators (with typical conversion efficiency of about 30%). The combustion temperatures of the conventional incinerators fuelled only by wastes are about 760°C in the furnace and in excess of 870°C in the secondary combustion chamber. These temperatures are needed to avoid odour resulting from incomplete combustion but are insufficient to burn or even melt glass. To avoid the deficiencies of conventional incinerators, some modern incinerators utilize higher temperatures up to 1650°C using supplementary fuel. This reduces the waste volume by 97% and converts metal and glass to ash. While incineration is extensively used as an important method of waste disposal, it is associated with some polluting discharges which are of environmental concern, although in varying degrees of severity. These discharges can, fortunately, be effectively controlled by installing suitable pollution control devices and by suitable furnace construction and control of the combustion process.

Thermal waste-to-energy technologies are the only solutions to deal with mixed wastes. In whatever way mixed wastes are treated, the impurities in them will pollute air, water, and land resources. By aerobically composting mixed wastes, the heavy metals and other impurities can be leached into the compost and distributed through the compost supply chain. In contrast, incineration is a point source pollution control technology, in which the impurities in the input mixed waste are captured using extensive pollution control technologies and can be handled separately. The bottom ash from incineration contains inert inorganic materials and is used as a construction material. The fly ash from incinerators can contain hazardous elements if air pollution controls are not properly designed or operated.

Only two incinerator plants have been built in India until now. The later of the two is situated at the Okhla landfill site, New Delhi. The previous incinerator, which was built in Timarpur, New Delhi, is not in operation anymore. Figure 6.5 shows a typical waste-to-energy diagram. As described in the figure, MSW can be used for mass burning without segregation. However, after mechanical segregation, an energy-rich fuel called RDF is obtained, which can be used to produce power through biochemical or thermal route. In the biochemical route, only anaerobic digestion has been used commercially, while in the thermal route, both pyrolysis and RDF burning have been commercially successful.

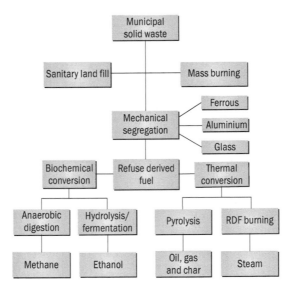

Figure 6.5 Options for energy production from municipal solid waste

A potential exists for generating an estimated 1500 MW of power from MSW in India. The potential is likely to increase further with economic development. The break-up of MSW generation and potential for energy recovery by state is given in Table 6.3.

Table 6.3 Potential for recovery of electrical energy from MSW by Indian states

State/union territory	Recovery potential (MW)	State/union territory	Recovery potential (MW)
Andhra Pradesh	107.0	Maharashtra	250.0

Contd...

Table 6.3 *Contd...*

State/union territory	Recovery potential (MW)	State/union territory	Recovery potential (MW)
Assam	6.0	Manipur	1.5
Bihar	67.0	Meghalaya	1.5
Chandigarh	5.0	Mizoram	1.0
Chhattisgarh	22.0	Odisha	19.0
Delhi	111.0	Puducherry	2.0
Gujarat	98.0	Punjab	39.0
Haryana	18.0	Rajasthan	53.0
Himachal Pradesh	1.0	Tamil Nadu	137.0
Jharkhand	8.0	Tripura	1.0
Karnataka	125.0	Uttar Pradesh	154.0
Kerala	32.0	Uttarakhand	4.0
Madhya Pradesh	68.0	West Bengal	126.0

6.6 DISASTER MANAGEMENT

Several disasters, both natural and man made, affect human populations around the world. These disasters damage property and take human and animal lives. These can be owing to the following events: earthquakes/tsunamis, floods, cyclones, snowstorms/avalanches, landslides, land subsidence/sinkholes, and droughts.

Disaster management involves a quick response to any of these disasters to minimize damage to property and reduce human suffering. This section will describe these disasters and the responses of governments and individuals for alleviating the results of the disasters.

6.6.1 Earthquakes

Earthquakes occur when there is a sudden movement along the faults (also called tectonic plates) below the earth's surface, either on land or in the ocean. There are several small earthquakes happening frequently in many parts of the world where there are active faults; however, they cause disasters only when they occur in populated areas and when the intensity of the earthquake is above 5 or 6 on the Richter scale (used to measure the intensity of earthquakes). Large magnitude earthquakes below the ocean's surface disturb the water surface and create large damaging waves called tsunamis. The 2004 tsunami destroyed property worth billions of dollars and killed thousands of people in several Asian countries, including India's southern coast. Japan has suffered a lot of tsunamis over the years.

In India, earthquakes happen along the Himalayan foothills and in several parts of the country where active faults are present. These have resulted in deaths and property damage. The following precautions need to be taken and preparedness is necessary to minimize loss of life and property damage due to earthquakes:

- Early warning programme based on field monitoring devices and a rugged and reliable communication system.
- Earthquake-resistant construction using materials and methods that will allow the concrete building or pavement to move but not collapse the structure.
- Planning of utilities in a manner that they are not destroyed during earthquakes, and any fires resulting from ruptured pipelines are easily controlled.
- Emergency preparedness procedures that will respond to the collapse of structures and other emergencies within the shortest time possible.
- Trained personnel in hospitals and government who can treat the injured and manage disruptions to the normal life in a community.

6.6.2 Floods

India faces floods every year with major rivers overflowing their banks. The monsoons bring heavy rains during a short period, and unless the flood plains of the rivers are preserved from development, major loss of property and lives occurs every year. Improved meteorological forecasting of the timing and intensity of monsoons will reduce the damage to property and loss of life from floods. Flood management, at a minimum, should consist of the following actions:

- Mapping and prevention of development in the flood plains of major rivers such as the Brahmaputra, Ganges, Yamuna, and Cauvery.
- Better control of storm water through the maintenance of green areas as well as rainwater harvesting through infiltration trenches/pits. Storage and use of rainwater can also solve the water shortage problems.

- Maintenance of drainage structures in urban areas and upgrading them to accommodate the expected intensity of storms.

The rivers in India can be broadly divided into the following four regions for studying the flood problem: (1) Brahmaputra region, (2) Ganga region, (3) north-west region, and (4) central India and Deccan region.

To monitor the possibility of floods, the Central Water Commission (CWC) has a flood forecasting system covering 62 major rivers in 13 states. There are 55 hydro-meteorological stations in the 62 river basins. The CWC monitors the water levels of 60 major reservoirs with weekly reports of reservoir levels and the corresponding capacity for the previous year and the average of the previous 10 years. Similar monitoring of smaller reservoirs by the irrigation departments of state governments gives advance warnings of hydrological droughts with below-average stream flows, cessation of stream flows, and decrease in soil moisture and groundwater levels.

6.6.2.1 General flood management measures practised in India

Different measures have been adopted to reduce losses and protect flood plains. Depending on the nature of work, flood protection and flood management measures may be broadly classified as follows:

- **Engineering/structural measures:** The engineering measures that bring relief to flood-prone areas by reducing flows and thereby flood levels include:
  - an artificially created reservoir behind a dam across a river;
  - a natural depression suitably improved and regulated, if necessary;
  - diversion of a part of the peak flow to another river or basin, where such diversion would not cause appreciable damage; and
  - construction of a parallel channel bypassing a particular town/ reach of the river prone to flooding.
- **Administrative/non-structural measures:** The administrative methods endeavour to mitigate flood damages by taking the following measures:
  - Facilitating timely evacuation of people and shifting their movable property to safer grounds by having advance warning of incoming flood, that is, flood forecasting and flood warning in the case of threatened inundation;

- Discouraging creation of valuable assets/settlement of people in areas subject to frequent flooding, that is, enforcing flood plain zoning regulation.

6.6.3 Cyclones, Hurricanes, and Typhoons

Cyclones, hurricanes, and typhoons are different names for the same type of storm. A tropical cyclone is called a hurricane in the North Atlantic Ocean, South Pacific Ocean, or North-east Pacific Ocean. A typhoon occurs in the Northwest Pacific Ocean west of the International Date Line. These storms have wind speeds of more than 74 miles per hour (119 km/h), and their intensity is rated on the Saffir–Simpson scale. Cyclones occur in the eastern coast of India and have caused extensive loss of life and property damage in states of Odisha and Andhra Pradesh.

On 4 October 2013, Phailin was the second strongest tropical cyclone that landed in Odisha, India, since the 1999 Odisha super cyclone. The system was first noted as a tropical depression on 4 October 2013 within the Gulf of Thailand to the west of Phnom Penh in Cambodia. Over the next few days, it moved westwards within an area of low to moderate vertical wind shear. As it passed over the Malay Peninsula, it moved out of the Western Pacific Basin on 6 October 2013. It emerged into the Andaman Sea during the next day and moved west–north–west into an improving environment for further development before the system was named Phailin on 9 October 2013, after it had developed into a cyclonic storm and passed over the Andaman and Nicobar Islands into the Bay of Bengal. Around 12 million people were affected.

The India Meteorological Department is responsible for cyclone tracking and providing warning to concerned user agencies. Cyclone tracking is done through the INSAT satellite and 10 cyclone detection radars. Warnings are issued to ports, fisheries, and aviation departments. The warning system provides for a cyclone alert of 48 h, and a cyclone warning of 24 h. There is a special Disaster Warning System for the dissemination of cyclone warning in local languages through INSAT to designated addresses in isolated places in coastal areas.

6.6.4 Snowstorms and Avalanches

Snowstorms cause economic disruptions and loss of life and damage to property every year in North America. In mountainous areas, snowstorms cause avalanches, which can be defined as a sudden slide of a large mass of snow on a mountain slope. With good meteorological data collection and analysis, prediction of snowstorms has improved and warnings are provided to communities. This helps manage the disasters resulting from snowstorms and avalanches. In Himachal Pradesh, Uttarakhand, and higher elevations in the north-eastern states, snowstorms and avalanches should be made part of the disaster management plan.

6.6.5 Landslides

Landslides are mass movement of soil on unstable slopes and are natural phenomena. However, man-made actions such as building houses on steep slopes where soils become unstable during heavy rains result in the loss of life and damage to property.

Nature's fury in Uttarakhand should open the government's eyes. What happened in Uttarakhand (Chaar Dham) was a calamity in all its ferocity, which unnerved the whole state/nation. The Himalayas were formed when the Indian landmass collided with the Eurasian landmass 55 million years ago. The process still continues, making the Himalayas rise. This triggers earthquakes, fracturing, shearing of rocks, and making slopes unstable and the mountains fragile, and vulnerable to human intervention and climate change. This results in the following:

- Hill slopes around a river bed get eroded.
- Existence of landslide-prone zones, including active landslide areas.
- Deadly flash flood-prone areas.
- Extremely weak zones of Bhagirathi and Alakananda rivers, which are flood-prone rivers.
- Stability of hill slopes depends on river flows and rainwater run-off causing landslide conditions.
- Active seismic and tectonic fault lines caused due to seismic movements in the Himalayas form weak planes.

Uttarakhand was once an environment-friendly abode, made sustainable owing to the fine ecological balance ensured by nature-loving, aesthetic, content, and tree-worshipping people. However, it seems the state has changed its track and resorted to cutting trees for building houses or farming, cutting mountains in the name of development, building roads at will, building dams to feed an unending greed, constructing hydro projects, and conducting mining operation.

Further, there has been a huge increase in the number of pilgrims, tourists, and expeditions, which has generated tonnes of rubbish and solid waste, dumping of plastic and waste in the holy rivers. The region's coping capacity has exceeded, causing adverse impact on the environment, including generation of greenhouse gases and rise in temperature.

To ensure sustainable development and ecological balance, all aspects should be assessed scientifically. Warming increases moisture-holding capacity and cooling causes cloud bursts. So the following points need to be ensured:

- Proper road alignment, avoiding weak plains, old debris-filled areas and so on.
- Drainage along roads and bridges as needed.
- Drainage in villages.
- Proper mining, if done scientifically, can reduce the risk of flooding. The mine plan should be in such a way that its bench alignment reduces the velocity of water during the monsoons when it refills the mine area. This will also avoid flash floods. Wrong planning and operations can cause flash floods.
- The river banks need to be protected at the micro hydroelectric plants at lower horizon, mining sites, sanitation facility away from the bank of rivers, flood plains, plantations all over, particularly on the banks of rivers.

It is extremely essential to have a cell to prepare a disaster management plan, including ecological atlas (containing information on vulnerable and sensitive areas), and a disaster management and warning system, for the whole region.

For preparing the plans, a cell consisting of representatives from the Geological Survey of India, Roorkee University, Indian Space Research Organization, Garhwal Vikas Nigam, metrological department, and experienced persons from the community should be formed. The focus should be on research and field studies to assess impacts of climate change, particularly on rains, in the region. An increase in temperature of 1°C will cause a variation up to 50% in rainfall and its pattern. Increase of temperature more than 2°C will invite hell.

6.6.6   Land Subsidence/Sinkholes

Land subsidence is the sudden collapse of the ground when the soil supporting the ground fails because of various reasons. Sinkholes are caused by the dissolution of limestone rock underlying the ground surface and commonly occur in areas with shallow limestone bedrock. Groundwater dissolves the limestone rock, and the ground suddenly collapses, killing people or damaging property. In India, land subsidence is primarily due to abandoned shallow underground mines which have not been backfilled with sand or other materials. Land subsidence is a disaster because it suddenly destabilizes structures on land. People living or working in these structures lose either their property or their lives or both.

6.6.7   Droughts

Droughts are major disasters around the world that cause food shortage and famine. These are usually human made due to the improper management of the water provided by nature. However, with global warming and unpredictable storms, droughts are happening even after good water management practices. Steps being taken in India to prevent droughts and manage water shortage include the following:

- Rainwater harvesting in all major urban areas.
- Management of storage reservoirs and implementation of water conservation measures.
- Wastewater recycling and reuse, especially greywater (from households) in urban areas and industrial wastewater recycling and reuse.

6.6.8   Disaster Management in India

For mitigating the effects of disasters by undertaking a holistic, coordinated, and prompt response, the Government of India established the National Disaster Management Authority (NDMA) in 2005. The NDMA has the following functions:

- Prepare policies on disaster management
- Approve national plan
- Approve plans prepared by the ministries or departments of the Government of India in accordance with the national plan
- Formulate guidelines to be followed by state authorities in drawing up the state plan
- Lay down guidelines to be followed by different ministries or departments of the Government of India for integrating the measures of disaster prevention and mitigation in their development plans and projects
- Coordinate the enforcement and implementation of the policy and plan for disaster management
- Recommend provisions of funds for the purpose of mitigation
- Provide such support to other countries affected by major disasters as may be determined by the Central Government
- Take similar measures for the prevention of disaster, mitigation, or preparedness and capacity building for dealing with the threatening disaster situation
- Lay down broad policies and guidelines for the functioning of the National Institute of Disaster Management

7

Social Issues and Public Participation to Minimize Impacts of Development

7.1 INTRODUCTION: EMBRACING SUSTAINABLE DEVELOPMENT FOR SOCIAL CHANGE

The need for development to address social issues is unquestionable. In India, roughly 70% of the population lives in poverty on less than ₹130 per day. But what type of economic development is needed in the country? Economic prosperity and high standards of living are typically equated with industrialization, yet industrialization is often blamed for the worst environmental issues. Many would argue that as India industrializes, social issues such as poverty, hunger, lack of water, poor sanitation, and diseases may decrease, but environmental issues will increase.

Fortunately, in the 21st century, industrial development can be achieved in a very different manner than the industrialization process of the 19th and 20th centuries. Today countries have the opportunity to engage in a more conscious type of development where traditional practices and lifestyles that honour the environment are maintained and also adopting emerging technology that provide jobs, food, sanitation, health care, transportation, and energy in the most sustainable ways. In a developing nation such as India, the contrast between traditional environmentally sustainable habits of living and unsustainable modern lifestyles is visible everywhere. People are learning about the ways of maintaining old with the new by increasing environmental education in schools and colleges. Other government entities, non-governmental organizations (NGOs), and corporations are also seeking to preserve, incorporate, and promote both traditional and newer concepts of sustainable economic development. Building rooftop solar panels are shown in Figure 7.1. The discussion, examples, and case studies presented in this chapter serve as the starting point for understanding successful and promising examples of economic development that address social issues and protect the environment.

Figure 7.1 Solar panels on a rooftop for a sustainable future
Picture courtesy Lynn Tiede

7.2 ENVIRONMENTAL ISSUES AND DEVELOPMENT: ENERGY AND WATER CONSERVATION, CONSUMERISM, AND WASTE DISPOSAL

India finds itself in the midst of a great struggle with environmentalism. Like all developing, primarily subsistence agriculture-based societies, amazing examples of connectedness to nature, which the fully industrialized Western world has lost, are found in India. In addition, deep spirituality in all its forms (Hinduism, Buddhism, Jainism, Sikhism, Islam, Christianity), Gandhian non-violence ideology, the more recent consumerist lifestyles that threaten these beliefs, and poverty have all commingled to create a culture very much in touch with the issue of sustainability.

An average American residing in New York City will consume far more than an Indian in his/her lifetime. It is easy to understand why. The average Indian uses 40–60 L of water a day, eats a more organic diet, recycles and reuses everything from paper to plastic, glass, bottles, and clothing. Recyclable wastes are sold to junk or scrap dealers. The incentive of getting paid for discarded items motivates middle-class Indians to utilize such services.

Packages and containers for serving homemade and street food in India are made from clay, newspaper, banana leaves, and metal, which can be easily composted, recycled, or reused. In many places, soda is still sold in glass bottles that are returned on the spot (Figure 7.2). Non-polluting motorcycles, bicycles (Figure 7.3), and rickshaws are used for transport. Rainwater harvesting is promoted and superfluous consumer goods are avoided. Many arduous tasks are performed manually.

People eat mainly unprocessed food. There is a can-do/make-do spirit that characterizes all daily activities. In India, the people live by the motto of improvising with the limited resources available. Such a lifestyle seems

Figure 7.2 Soda served in glass bottles, which are returned on the spot, encouraging reuse of glass and reduction in trash reaching landfills
Picture courtesy Lynn Tiede

Figure 7.3 Bicycles used for commuting
Picture courtesy Lynn Tiede

like a utopia to an environmentally friendly advocate of sustainable living from the West.

Environmental concerns arise when one realizes that the sustainable way of life is changing with industrialization. The Green Revolution in the 1960s, which brought pesticides and fertilizers to agriculture, could be viewed as the first step in this process since India's independence from colonialism.

Multinational corporations entered the Indian market in 1991 when the government opened the market for foreign investments. The country proved to be an ideal location for multinational businesses to set up shop due to cheap labour, few enforced environmental regulations, and a largely English-speaking population. In this growing economy, technologically trained and highly educated middle- and upper-class Indians have more disposable income and, as a result, foster consumerism. Large cars, motorcycles, packaged goods, and shopping malls have all arrived in India. Media also contributes in promoting a consumer-oriented lifestyle.

These developments are occurring in the context of a young 68-year old democracy. However, stronger efforts need to be made to protect India's natural environment. For businesses, adherence to international environmental standards is voluntary, and India's citizens are not fully organized in putting pressure for respecting environmentally sustainable traditions. People are distracted from protesting against environmental issues because of the demands of daily life and survival. For instance, all over India groundwater is getting contaminated by agricultural pesticide-laden run-off and is being depleted by industries, hotels, and growing population. When Centre for Science and Environment (CSE) investigated bottled water and carbonated beverages, it found that these beverages not only contain high levels of pesticides but also utilize huge quantities of water in their manufacturing. Indians are using 6540 million bottles a year, that is, about six bottles per person per year. In Delhi, it is an average of 50 bottles per person per year. The Indian Government is yet to enact stricter regulations to monitor this problem, and beverage companies have criticized CSE for its work.

Besides such water issues, there are other environmental concerns in the face of development and social pressures. Population growth, demands on resources due to large industries in small villages, and the difficulty in subsisting on small-scale agriculture have pushed millions of poor people to urban centres, creating shanty slums (Figure 7.4). These slums accompany unsanitary conditions, contaminated water, and

Figure 7.4   Unsanitary conditions of urban slum areas
Picture courtesy Lynn Tiede

breeding grounds for diseases in the uncollected trash piles. Slum dwellers prepare food on the street in the midst of these conditions. In addition, stray cattle create other health hazards as well.

Despite India's rich history of environmentally sustainable practices and habits, in today's time, environmentalists have to find solutions to the current environmental and social problems and promote sustainability within the new phase of economic development. Part of this challenge of resolving India's environmental and social issues (and not creating more) is realizing that there is no all-encompassing solution. Requiring environmental education at every level, from primary school through university, would allow the average citizen to gain an understanding of the need for sustainable economic development and gradually bring this knowledge into everyday living, career choice, and voting and advocacy efforts. However, finding solutions and perhaps holding on to the good past habits must be immediate in a country that is rapidly industrializing.

In India, building of environmental awareness is supported by entities outside of mainstream education. Sustainable economic development aiming to foster both environmental protection and respect for human rights, is occurring via various stakeholders through political and educational initiatives and reforms in business practices. The Indian Government, many NGOs, and even corporates are leading the way. They are directly taking action to address issues in the present. Learning from the mistakes of Western industrialization in the 1800s and 1900s, India's largest corporations and non-profit organizations have developed some of the most innovative large-scale approaches to create a green society and economy in the world. This highlights the awareness of business and government leaders towards India's need and the exciting opportunity to fully develop and industrialize in a sustainable way. These leaders are ensuring that India's traditions of environmental sensitivity and sustainable use of energy, water, and natural resources of the old generations will continue alongside vibrant economic development, which provides a basic standard of living for India's citizens.

7.3 GOVERNMENT INVOLVEMENT IN PROTECTING THE ENVIRONMENT

7.3.1 Environmental Legislations

The most obvious way through which a democratic government protects various rights is making legislations. India has established many regulations to help protect the environment since the 1970s.

India was the first nation to make environmental protection an amendment to its constitution after the UN Conference of the Human Environment in Stockholm in 1972. The conference galvanized many nations of the world to begin addressing environmental protection at the dawn of the modern environmental movement. India's regulations govern the legal responsibilities of the government officials and agencies towards the environment when considering government actions. The Constitution of India also makes it a duty of every citizen to "protect and improve the natural environment including forests, lakes, and rivers, and wildlife." The acts establish a central monitoring body, the Ministry of Environment, Forests, and Climate Change (MoEFCC), and boards such as the Central Pollution Control Board (CPCB) to monitor and regulate key legislations and delineate penalties for violation of the acts. These and multiple other existing laws and guidelines in India's Constitution, if effectively implemented and enforced, can do much to protect the environment. These regulations also provide a foundation for legally challenging the actions of other entities and evaluating whether an issue of concern is legally protected. They may also provide impetus for the creation of more legislations to protect elements of the environment. Some key acts are summarized in the following sections.

7.3.1.1 *Environmental Protection Act, 1986*

This act provides for the protection and improvement of environment and for matters connected therewith. It relates to the protection and improvement of the environment and the prevention of hazards to human beings, other living creatures, plants, and property. It includes the very important Hazardous Wastes (Management and Handling) Rules, 1989; the Manufacture, Storage, and Import of Hazardous Chemical Rules, 1989; and the rules for the manufacture, use, import, export, and storage of hazardous microorganisms, genetically engineered organisms, and cells.

7.3.1.2 *Air (Prevention and Control of Pollution) Act, 1981*

This act provides for the prevention, control, and abatement of air pollution, for the establishment of boards with a view to carry out the aforesaid purposes, for conferring on and assigning to such boards powers and functions relating thereto and for matters connected therewith.

7.3.1.3 *Water (Prevention and Control of Pollution) Act, 1974*

This act provides for the levying and collection of cess on the water consumed by certain industries and by local authorities with a view to augment the resources of the central and state pollution control boards for the prevention and control of water pollution.

7.3.1.4 Wildlife Protection Act, 1972

This act is a comprehensive national law with the sole aim of protecting wildlife and plants and for matters connected thereto or ancillary or incidental to, with a view to ensure the ecological and environmental security of the country.

7.3.1.5 Forest Conservation Act, 1980

This act, with a precursor passed in 1927, is designed to halt further deforestation and to conserve forests.

7.3.2 Government Initiatives and Partnerships

The governmental initiatives cited in this section provide the starting point for understanding how elected officials and state workers are facilitating broader holistic efforts to ensure that development projects throughout India are environmentally sustainable and adhere to legislated directives.

The Government of India, though criticized at times for not placing stringent environmental regulations or enforcement policies on businesses, does promote sustainability through the Planning Commission's five-year plans, the Ministry of Industry's Cleaner Production Centre initiatives, and the Ministry of New and Renewable Energy. The government promotes the 'Ecomark' label under the CPCB. This somewhat struggling project, implemented in 1991, is an effort to label environmentally friendly products for consumers and motivate businesses to gain the label.

The Indian Government is also partnering with other nations. In 1997, the Clean Technology Initiative (CTI) was started as a joint effort of the United States Agency for International Development (USAID) and the Industrial Credit and Investment Corporation of India (ICICI) Bank, to promote climate friendly technologies and certifiable environmental management systems. The Taj Trapezium Zone was started under this initiative to reduce the destructive influences of industry on Taj Mahal and 40 other monuments. The project was designed to promote cleaner production processes in glass, bangles, and diesel generator manufacturing. It focused on creating awareness of clean technologies and conducting factory walkthroughs to identify cost-effective and energy-efficient options. It facilitated obtaining necessary financing to render the green chain supply. The CTI has also provided other greening assistance to many industry sectors across India.

The US–India Climate Change Partnership is also a joint project between the US Department of State and the Government of India's

MoEFCC. One arm of this initiative, the Greenhouse Gas Pollution Prevention Project, is working to reduce greenhouse gas emissions of thermal power plants. It is also working to help sugar mills to utilize biomass for energy needs, using highly efficient cogeneration units. In addition, the project is funding and promoting the following:

- Solar, wind, hydro, and biomass technologies around the county
- Development of Reva, India's first electric car (Figure 7.5)
- Sponsorship of an eco-hotel chain called the Orchid Group of Hotels
- Suitable reuse options for fly ash (ash produced during combustion of coal)
- General improvements in energy supply and distribution
- Technical assistance to India to develop hydrogen as a fuel, including the development of a hydrogen-powered three-wheeled scooter

The US Asia Environmental Partnership brought Ajay Singh, the Chief Environmental and Sustainability Officer of the New York City Transit (NYCT) to New Delhi. Singh helped the NYCT receive an ISO 14001 certification and helped the New Delhi Metro System achieve this rating in the construction phase. This makes them the only two transit systems with this rating. Singh incorporated management practices that consume less paper and power, stations that use solar panels, and recycling close to 85% of construction waste. Thus, the Government of India has played a role in forging India's path to sustainable development.

Figure 7.5 Reva electric car
Picture courtesy Lynn Tiede

7.4 NON-GOVERNMENTAL AND CORPORATE INVOLVEMENT IN PROTECTING THE ENVIRONMENT

7.4.1 Environmental Ethics of Corporate Social Responsibility and Environmental Responsibility

Corporate social responsibility (CSR) is a term widely used by businesses and NGO's alike. While there are conflicting definitions and understandings of the term CSR, its use started in the 1980s when NGOs such as Greenpeace began protesting by advocating boycotts, shareholder action, ethical shopping, direct action, and media campaigns to draw attention to environmental catastrophes such as Exxon's oil spill in Valdez, Alaska, USA, and Nike's abusive child labour practices.

With these organizations in the lead, citizens are changing from consumer citizens to citizen consumers. As multinational corporations grew in power and size and spread, concerned citizens began acting as watchdogs in the absence of international organizations setting standards for operations in countries where environmental and labour laws were weak. Companies soon realized that they had a responsibility to all their stakeholders (consumers, community members, government officials, employees, environment, future generations, and shareholders) and that consumers were demanding ethical values and caring for people, communities, and the environment.

Not only was CSR responding to consumer demand, acting and providing proof of such commitment could favourably affect a company's image and perhaps increase profitability. Socially and environmentally concerned heads of companies saw CSR as a way to redefine capitalism and make it work as a tool to promote a better quality of life and sustain business for the future generations. Environmentally responsible initiatives such as improving energy efficiency of production equipment could save money and increase profits as well. CSR made good sense to many business leaders on all these levels.

Over the last 15 years, efforts to incorporate responsibility have grown in many directions. CSR can include the following:

- Environmental responsibility in operations (water management, energy efficiency, zero waste)
- Environmental initiatives in communities where located (tree planting, sanitation projects)
- Assessing environmental and social impacts before starting a new project

- Manufacturing products and providing service in environmentally friendly ways
- Developing products that are safe for both consumers and the environment
- Developing products that can be recycled, reused, or disposed of safely with no additional environmental contamination
- Promoting environmental and social responsibility with contractors and small-scale suppliers
- Communicating and fostering openness with employees and consumers about potential hazards of business
- Promoting higher levels of environmental and labour performance than those legislated in the country of operation
- Labour rights (wages, working conditions, hours of employment, union organizing)
- Advancement of women's rights and women in the workplace
- Development of corporate-run initiatives to improve the economic livelihood and alleviate poverty of people in the community where located
- Empower local communities in decision-making and economic development planning
- Create educational opportunities in communities where located
- Improve infrastructure in communities where located
- Understand individual community needs
- Address health concerns such as HIV/AIDS
- Address population growth
- Respect animal rights
- Contribute financially to NGOs involved in the above issues

International organizations, such as the World Business Organization, International Standards Organization, and United Nations, have encouraged sustainable economic development and CSR, and linked these to the survival of human beings. These organizations believe that if standards of living are to improve for the poorest and not worsen for those in industrialized nations, businesses and consumers must live and conduct their work in ways that promote reverence for life and the conservation of natural resources.

In India, large corporations, such as the ITC Group of Hotels, Tata Steel, Jubilant Organosys, Sahara India Pariwar, ChemFab Alkalis, and the Murugappa Group, are working to incorporate comprehensive CSR or sustainability models in their operations. In other words, they are striving to look at how well the businesses are doing in maintaining the three bottom lines—profit, people, and communities—and the environment.

It is difficult to assess the sustainability and corporate responsibility of many small and medium enterprises (SMEs), which make up 45% of India's economy. Many of these companies practise sustainable and environmentally friendly business models, but they are not certified as such. Many fair-trade cooperatives organize and promote women's handicrafts and a growing number of organic farms clearly operate with ethical, environmentally, and people friendly principles in mind. TARA Projects, Biotique, Sadhna, Fair Trade Forum India, and FAB India are examples of these.

Large corporations and organizations talk about greening the supply chain and are working to encourage multinationals that subcontract work to smaller businesses to teach SMEs sustainability principles and to identify and highlight the ways in which they are already engaging in such practices. Helping SMEs develop capacity and gain financial backing and access to markets through partnerships with large and multinational corporations would seem a way forward for all.

There is a growing concern that in many businesses, CSR has become 'green washing' and is done purely for building a positive image in the eyes of the consumers. It is not seen as genuine or comprehensive enough efforts. For example, although beverage companies have challenged CSE's findings of pesticides in their products, they sponsor environmental education in schools. Thus, at times CSR may not be responsive to the actual community's needs and becomes more philanthropic, rather than challenging or reformulating the structure and operation of that business in the community or country. It may not go far enough in challenging profit-oriented businesses to examine their practices and goals as truly respectful to human life and environmentally sustainable on the earth. This is a challenge all over the world and not just in India. However, though more depth of integrity in practice is still needed, CSR is an important concept for all businesses of the world to embrace. While industrialization and corporations often get blamed for environmental destruction, understanding how businesses are implementing CSR and sustainability on the ground provides insight for those seeking to green the world. Like the precautionary principle (discussed in Chapter 1), CSR signifies a pathway towards living in harmony with nature as the future unfolds.

Some general CSR initiatives in the areas of energy and water conservation, recycling, and reduction in pesticide usage in farming provide the initial examples.

7.4.2 Sustainable Energy Initiatives

As oil prices increases and supplies dwindle, India is critically aware that it must find new sources of energy to fuel its demand. Coal is used to generate 52% of India's energy but is recognized as being unsustainable, with 100–120 million tonnes of fly ash being produced. Compressed natural gas (CNG) has replaced diesel and unleaded fuel in many parts of the country. It has reduced particulate matter and other toxics. Energy-saving technologies are being promoted and developed for industry. Organizations, such as the Society for Research Initiative in Sustainable Technologies and Institutions, Auroville, and the Gujarat Grassroots Innovations Augmentation Network are working to promote grassroots developed green technology. An example is a water cooler run by the heat of evaporating water, developed by a villager in Gujarat.

Solar power is underutilized in a nation that gets 10 months of sun. Other renewable sources, such as biomass, water, wind, geothermal energy, and hydrogen are other possibilities. Energy efficiency within industry and automobiles, as well as better mass transportation, can help improve the situation.

According to the CSE's Green Rating Project, the automobile industry has a long way to go in terms of improving corporate environmental policy and management systems, corporate leadership and proactive environmental initiatives, procurement policy and supply chain management, process and consumption efficiency, pollution and pollution control and prevention, product use, and product disposal. In addition, focus on design for engine efficiency and utilization of the cleanest fuels possible will push the industry towards sustainability. Reva, jointly produced by the Maini Group (Bengaluru) and Amerigon (Monrovia, California) with the support of USAID, is India's first electric and battery-run vehicle and has been doing well in the country and in the world market.

The Muragappa Group (Parry Sugar Industries Ltd) is an innovative company that uses multiple resource recovery, cogeneration, and ethanol production to operate a highly sustainable sugar-manufacturing company. While the company is the oldest in India and was originally started by the British in 1842, it is now quietly transforming into a green business with multiple CSR components. As is true in much of India, such transformations are not always labelled or certified green, organic, and fair trade, but they are happening nonetheless.

Parry Industries' Nellikuppam sugar factory in Tamil Nadu is supplied by a cooperative of 20,000 farmers. Farmers receive support services and training in integrated pest management, composting, mulching, drip irrigation, and other less chemically and resource-intensive methods of farming. Social, medical, general education, and computer training services are also provided to farmers in the local villages. The company works to make sure electricity and healthy sanitation systems exist in the communities where its farmers work. Parry takes pride in knowing that farmers are being paid a fair wage, and that communities are benefiting economically, thus alleviating poverty in this part of the world.

Energy efficiency and environmental responsibility are also priorities for Parry. Sugar production is an energy-intensive process, but cutting-edge, energy-efficient cogeneration technology is used throughout the factory. Sugar cane waste is burnt in the factory to produce energy to run the mill, but much of it is recaptured. This allows Parry to supply energy to India's grid. The liquefied waste molasses is utilized at a distillery to create both spirit alcohol and ethanol, both of which are sold to other producers. The wastewater effluent from the distillery is treated at a plant with anaerobic and aerobic digesters to produce highly organic manure (Figure 7.6). Biogas is captured from the process as well, which then creates steam to power the distillery. Rainwater is also harvested at the plant, and there is zero discharge. Other products such as sugar cane bagasse, molasses, and press mud are distilled and sold as liquid and solid cake-style fertilizer/manure for sugar cane farmers. Since farmers pay very little for it, their expenses remain low and the use of chemical fertilizers is reduced. The company has an energy manager and energy auditors on staff to encourage constant exploration of new technologies and better ways to implement energy-efficient and environmentally sustainable practices.

Another non-profit organization is The Energy and Resources Institute (TERI). Headquartered in New Delhi, TERI seeks to educate and improve energy efficiency in Indian industries through research, consultancy, training, and information dissemination. Its mission and scope of work

Figure 7.6 CSR initiatives by Parry Industries
Picture courtesy Lynn Tiede

is towards encouraging sustainable development on an international level. The institute is itself a green building and takes pride in bringing renewable energy to rural communities as well as providing educational programmes for children and teachers.

7.4.3 Water Conservation, Rainwater Harvesting, and Watershed Management

India is facing acute water crisis. Domestic usage of water is about 5%, agriculture usage is about 82%, and industry usage is 8%–10%, which is expected to double in the next 10 years. Industrial usage is a double-edged sword because it not only depletes groundwater sources but also creates highly polluted wastewater. This combines with the already taxed water systems, full of pollutants from pesticide- and fertilizer-laden agricultural run-off.

From 1998 onwards, the ancient practice of rainwater harvesting was revived and utilized in the country as a means of addressing the water shortage. Rainwater harvesting is the process whereby rainwater is collected from rooftops and other surfaces and routed underground so that the water filters naturally and replenishes the water table (Figure 7.7). Otherwise, as happens in most built environments, the water runs off through drainage systems into larger bodies of water, leaving groundwater aquifers depleted. Rainwater harvesting is mandatory in some places in India, and many municipalities, schools, and businesses have initiated such practices on their own.

Figure 7.7 A rainwater harvesting tank
Picture courtesy Lynn Tiede

7.4.4 Waste Management and Recycling

What happens to trash in India? First of all, most Indians live in a way that does not create much trash in their homes, although increasing consumption of Western-style consumer products with excessive packaging has changed this. There is an informal system of recycling run by kabadiwalas (scrap dealers), who frequent localities collecting discarded objects from houses and shops. They take paper, plastic, glass, metal, electronic appliances, cloth, shoes, 'anything and everything'.

Kabadiwalas dismantle objects if necessary and market the usable portions to larger buyers. Households are paid for the things they give to kabadiwalas, a pay-as-your-throw incentive. Organic and compostable trash is tossed into large holding areas on roadsides or near apartment complexes and is removed by municipal trunks. Rag pickers can be seen sorting through these piles to salvage anything missed by kabadiwalas.

It is estimated that kabadiwalas take 40% of plastics and recycle them, although polybags are a hazard as these are not easily recycled. Stray cows and dogs chew on the bags as they try to eat the leftover food inside. These bags do not disintegrate naturally and clog waterways. Plastic water bottles are also a new menace and are not easily recyclable. For this reason, many municipalities are working to ban plastic bags and bottles, and returning to using natural materials such as newspaper, jute, and cloth bags. The state governments are banning the use of plastic bags under the Environment Protection Act, 1986.

India produces large quantities of paper, both in large-scale paper mills and in small handmade paper businesses. Surprisingly, except for the small initiatives that source from straw, grasses, and waste rags, India primarily uses wood and bamboo as the major raw material. Wastepaper makes up only 9% of India's new paper production resources. Hence, there is large potential for collaboration with regular kabadiwalas and thiawalas (workers who collect wastepaper during office hours). Currently, much of the wastepaper they source is being put on the global market for wastepaper. Water misuse and pollution by paper companies, especially those that bleach paper, present other environmental challenges that need to be addressed.

Three other concerning areas of waste management are e-waste, organic matter, and fly ash. E-waste is a new concern, and other than the informal sectors that dismantle, rebuild, and reuse electronic components, the Indian Information Technology (IT) industry has just begun to address ways to combat this growing source of waste. India has been the dumping ground for outdated computers by developed nations apart from the country generating its own sizeable amount of e-waste. IT industries are advocating for producer responsibility towards the creation and disposal of e-waste. A large nation like India has particular challenges, although it already has much extended lifecycle for most products. Linking informal and formal recycling systems seems to be the answer.

Indian law requires that all organic matter be composted or converted to biogas, and this is slowly being implemented in various parts of the country. Proponents of organic farming have suggested that this is an untapped resource for fertilizer. Indeed, the ITC Hotel chain has incorporated composting at all its hotels, and other businesses and non-profits are also conducting small localized efforts. However, no national plan is in place to compost trash in all neighbourhoods.

Fly ash, 108 million tonnes of it, ends up in ash ponds and landfills. The Government of India has made the use of fly ash (25% by weight) mandatory in making clay bricks, tiles, and blocks by companies located within 100 km of coal- or lignite-based thermal power plants. More pressure and regulation to reuse this material in cement would reduce contamination of soil and water and decrease carbon dioxide produced in making cement.

Chintan Environmental Research and Action Group, India is an organization working to unionize rag pickers and promote their dignity, who work to reclaim natural resources from the waste stream. Chintan helps them get government-issued identity cards to validate their work and teaches them safety procedures such as using tools, gloves, and masks during their work (Figure 7.8). The organization also wants to end harassment by residents and police as well as help waste pickers

Figure 7.8 Chintan workers separating trash
Picture courtesy Lynn Tiede

have access to medical care and education and slow the privatization of waste collection by large corporations.

7.4.5 Organic Farming: Food/Cotton

Organic farming is an age-old practice in India. Farmers are still utilizing ancient techniques passed on by their forefathers. The much-lauded Green Revolution in India included high usage of pesticides. In 2003, India set up the National Board for Organic Production to address the need to formally enter this profitable $37 billion global market. There is the Ecomark, a stamp for organic and environmentally friendly production processes, and 'India Organic' designation for farms and products meeting required standards. India has shown growth in organic production in some commodities. Problems in organic farming, include high expenses for getting certified (though many are indeed organic, they cannot get the stamp of approval), high labour requirements, and poor marketing channels. There is also a period of transition when farmers move into organic farming and crop yields are less, but higher prices are not possible because certification requires the land to be chemical free for a number of years before products enter the market as organic. Many farmers cannot afford this transition. Overcoming these barriers seem critical for a path to a sustainable future.

Another hindrance in moving towards organic farming is the sickeningly high levels of pesticides in the water supplies and, therefore, food chain. Pesticide use in agriculture started in 1949. In 1954, the first plant to produce dichlorodiphenyltrichloroethane (DDT) and benzene hexachloride (BHC) began operations. India is the fourth largest pesticide producer in the world. While India's per capita consumption of pesticides is low, India uses the most toxic insecticides in the world, which make up 80% of all consumption. More than 50% of the chemicals are used in cotton production. Farmers have been committing suicide because of the debt brought on by chemical farming. Organic farming presents a hopeful alternative, especially if the government works to include poor farmers and not just subsidize large factory farms.

Organic India Pvt. Ltd is a well-known international company that has been certified organic by the United States Department of Agriculture and European Union, among others. They sell tea, spices, herbs, and other food products. They have been in operation since the 1990s and are promoting organic farming, and sustainable development.

The internationally renowned Vandana Shiva and her NGO have a similar mission. The organization she founded, Navdanya, works to educate citizens about the challenges of genetic engineering, biopiracy, food, water, and land sovereignty, and to preserve native plants and seeds. It runs an educational centre, hosts lectures, and has a seed bank. Navdanya challenges the philosophy of green revolution and globalization, and is involved on a global level actively lobbying for ways to protect the rights of native people to utilize their own knowledge and indigenous technology and techniques to live in dignity while protecting the environment. The goal is to bring all citizens of the world back into a lifestyle and reality, that cherishes and respects connectedness to each other and the earth. By promoting and celebrating organic and sustainable farming practices, Navdanya hopes to help citizens reconnect to truly sustainable ways of living, which include an emphasis on ecological responsibility and economic justice versus greed, consumerism, and competition as objectives of human life.

All businesses and NGOs seeking to follow CSR principles seem to be pursuing this world view and symbolize ways that ordinary people are using entrepreneurship and a sense of justice to promote environmental protection and a way forward for India's economic and social development. The following case studies provide a few more in-depth examples of this work.

7.5 CASE STUDIES IN EFFORTS TO PROMOTE SUSTAINABILITY

7.5.1 Non-Governmental Organizations

7.5.1.1 *Centre for Science and Environment*

The Centre for Science and Environment is one of the most well-known environmental advocacy groups in India. Its chief endeavour is the magazine *Down To Earth*. The organization helps frame and redefine the debate on issues and challenges the government and citizenry to examine all environmental concerns and start campaigns to make necessary changes. Its viewpoints are always backed by scientific analysis, which makes their arguments strong.

The CSE also has a programme titled 'Green Rating Project'. It has rated the paper, mining, automobile, and caustic-chloride industries in an effort to point out areas where environmental and corporate responsibility could improve. In addition, it offers regular training courses to improve environmental management in government and industry. CSE's research, campaigns, and green rating system have prompted companies to increase their environmental sustainability.

The environmental education department of the CSE works with children and teachers. It publishes *Gobar Times*, a supplement of *Down To Earth* for students and educators.

It created a green rating programme, in December 2004, to assess and promote sustainable operational procedures in schools. Both the magazine and the rating project encourage children and teachers to collectively work towards the goal of environmental sustainability at the local level. The CSE has also created the Anil Agarwal Environment Training Institute to communicate the science, complexity, and politics of environment across India, South Asia, and the world, and to build a constituency and cadre of knowledgeable, skilled, and committed environmentalists.

7.5.1.2 *Confederation of Indian Industries*

The Confederation of Indian Industries (CII) is India's oldest business organization. It works to shape policies that are acceptable to both businesses and the government.

The environmental management division of CII provides technical assistance to companies to help them implement ISO operational and safety standards and addresses other specific environmental problems. This division was created after the 1992 United Nations Conference on the Environment in Rio de Janeiro. Many companies and industries have much room for improvement, many of which are affiliated with multinationals, including prominent American ones. CII conducts training to encourage businesses to meet ISO standards. ISO 14001 deals with environmental standards, and ISO 18001 deals with fair labour practices. CII assists with technical help to improve operations in these areas.

The organization is also a resource for information. It relays information to and from the government and businesses and other organizations to promote competitiveness, growth, and sustainability of Indian businesses. As part of this information-sharing initiative, its environmental division has created a Green Building Centre in Hyderabad, the first Leadership in Energy and Environmental Design (LEED) platinum-rated building outside the USA. This centre assists companies wishing to go green with technical support and is a resource for the latest innovations in green technology.

7.5.2 Corporate Case Study

7.5.2.1 *ITC Welcome Group Hotels' comprehensive environmental initiative*

Individual companies are taking giant steps to encourage sustainability within their operations. As noted previously, the CII has trained many companies in ISO 140001, 180001, and SA 8000 standards. With the help of the Canadian Government, it trained businesses such as

Tata Steel, Jubilant Organosys, and Sahara India Pariwar in corporate sustainability management. The project was designed to raise the awareness of environmental issues and reduce the environmental impact of Indian industry on the environment using management systems. One company that took extremely comprehensive steps to promote CSR and environmental sustainability is ITC's Welcome Group five-star hotel chain.

It is headquartered in the platinum-rated green building in Gurgaon, Haryana. The building utilizes many green technologies. Solar energy is used to heat water; rainwater is harvested; a sewage treatment plant on site allows reuse of greywater; energy use is 50% less than a typical building; water use is 40% less; and natural daylight is used for lighting. ITC teaches others about the possibilities of such architecture by opening the building once a week and using touchscreen display to highlight the building's green features (Figure 7.9).

Besides housing itself in a green building, Welcome Group's environmental initiatives seek to make its hotels ecological pioneers of the country. The group continually works to go beyond mere compliance and stresses on the four 'Rs'—reduce, reuse, recycle, and rethink. Its management is always looking for ways to reduce wastes at all levels. It operates in environmentally friendly ways, trains employees to understand this philosophy, and works to train its local suppliers (the supply chain) to operate in the same environmentally friendly ways. It shares its best practices and guidelines on sustainability with SMEs through full-day training sessions. The training includes seeing green initiatives in action at the hotels as well as site visits to SMEs to analyse and show specific things they can do to improve their operations.

Figure 7.9 Signs proclaiming ITC Hotel's commitment to environmental sustainability
Picture courtesy Lynn Tiede

The initiative also works to teach hotel guests the importance of environmental issues through signs and information in the rooms and the lobby. It does environmental clean-up and plants trees in local communities. The company has designed an educational programme for teachers and students and is working to help green schools as well. Today this chain of five-star hotels throughout the country continues to incorporate gardening, composting, rainwater harvesting, greywater reuse, waste reduction, recycling, energy efficiency measures (sensors on lighting, asking guests to voluntarily reduce use of sheets, towels), and the use of non-toxic cleaners.

7.6 PUBLIC PARTICIPATION IN PROTECTING THE ENVIRONMENT

Public participation is very important in promoting environmental sustainability in the face of economic development. The democratic principles that enable such participation must be understood and put into action if solutions to environmental issues are to be found. Case studies illustrate the process and realities of civic involvement around environmental issues.

7.6.1 Democracy and Environmental Stewardship

Many believe in the democratic form of government because of the rights that citizens are given under the government. While enjoying life, liberty, equality, and pursuit of happiness, it is also important to take seriously one's responsibilities as a citizen in a democracy. What are the responsibilities? Based on one's time, education, and interest, the list might include paying taxes, obeying laws, staying informed, voting, involving in community decision-making, volunteering to solve problems that affect and concern people, and running for public office.

Lack of citizen engagement can have adverse effects. In a democracy, one of the most fundamental and important rights citizens have is to voice their opinion about issues affecting their communities. Without the involvement and perspective of many, a democracy fails in essence. If citizens understand the importance of this, they often voice opinions and get politically involved during electing the government who will make day-to-day decisions that may affect the community. Some civic-minded people can and do remain involved in local community boards and come to periodic meetings to voice opinions on all sorts of issues, including government projects planned for their community. Others write letters and relay concerns to elected officials when legislative issues are at hand. But on the whole, finding time for community involvement often seems like a luxury to those juggling work, family, and other life responsibilities in the modern world.

However, it is crucial to begin finding ways to engage. Poverty, overcrowding, habits of materialism, overconsumption, pollution, environmental degradation, and alienation from nature and community are threats to plants, animals, and human life on earth. While Western nations have driven these ways of the world via industrialized capitalistic and consumer-focused economic development, they are encompassing the globe as all nations 'modernize'. Without the efforts of all, these problems will never be solved and may worsen. It is especially important that all citizens and viewpoints be heard when considering economic development projects. Such projects present opportunities for deep reflection and creativity. What can emerge are projects that truly integrate life-giving values, which sustain humanity and the earth.

Most private development projects, whether they create new industries, shopping centres, housing or hotel properties, cannot gain full approval for siting without an environmental review process. All localities have their own particular procedure for such a review, but all democratic governments do allow for public input in the process. Citizens can often, by law, write or publicly (at specified times) voice their opinions. In all democracies, the rights to free expression and to petition the government for a redress of grievances are protected. Educated and environmentally conscious citizens thus have a right and responsibility for involvement in public affairs. This is critical when development is happening at such a rapid pace. It is easy to brush over stringent environmental concerns and even regulations, especially if there are promises of new employment for the poor, huge profits to be gained, or if there is corruption among government entities involved in approving the project. Without outcry of citizens or NGOs, development can happen in a helter skelter fashion, can lead to economic gain for a few, and can end up causing extreme environmental harm to a community.

7.6.2 Education

The first step in proactive citizenry is education. Citizens must be educated on many fronts if they wish to become involved in the decision-making process of their community. Local citizens often intuitively understand the threat of a particular project in their community. Such basic, from the ground up, day-to-day knowledge can be tapped upon to provide strong arguments for or against a project. However, it is important to develop one's argument and background before engaging in what can become a very emotional public participation. One will be better able to find the healthiest consensus and truly provide thoughtful alternatives or improvements to a proposed project. Otherwise very serious and important concerns get disregarded.

Some essentials to understand in this education phase are as follows:

- **Political process:** What are the particulars of the political process in one's locale? How and when are projects submitted for review? At what point can citizens provide input? What types of input are most valued or respected or responded to by government officials and the entities promoting the project—written, verbal, or more direct, creative protest methods? Can citizens make direct input to the developer or must they go through government officials or certain representatives of the body proposing the project?

- **Legal protections:** Are the citizens' concerns protected by law? What laws, rights, and legal precedents might be utilized to support various concerns? Are these legal protections recognized at the local, state, national, and/or international levels?

- **Community needs assessment:** What are the economic, social, infrastructure, health, and environmental needs of the community? What is the priority of each? How does each area interconnect and relate to the other? How can these needs be balanced? Who will be benefitted most from this project?

- **Environmental impact:** How will a particular project impact the environment in the community? Consider impacts on air, water, noise quality, health, safety, waste management, natural habitats, resources, and animals. Will water and electrical use of the proposed project deplete or interfere with what is available for others locally? What waste will be created and how will it be disposed of? What other natural resources will be utilized in the community to help this entity operate? How will this depletion of natural resources affect the community?

- **Considering viewpoints:** Are there other groups involved or concerned about this project? What are their concerns? What is their point of view? How can joint efforts make the project more environmentally sustainable and build community? How can various viewpoints clarify all facets of the project's impact? Are there any groups or NGOs who could get involved to support or provide research on this issue?

7.6.3 Building Consensus

Generally, when a new development project is proposed, the benefits of the project are presented first. More jobs, more prestige or modernization, more tourists, better quality of housing, life, and more convenience are promoted as the overriding concerns that make the project seem a win-win for all. Clearly these benefits are not inconsequential. From a sustainability perspective, however, the holistic way the project affects other facets of life in a community and the environment must be considered. Often, with discussion, multiple points of view, expert advice, reflection over time, and the willingness on the part of all entities to compromise, a truly sustainable, environmentally friendly solution can be found.

In a business, all parties affected by a project are called 'stakeholders'. The belief that all stakeholders—whether company owners, workers, local citizens, the government, or even silent stakeholders such as wildlife and ecosystems—should have a say in how and where a business is operated is essentially a belief in consensus building. In today's green-oriented world, the more sensitive to environmental issues all these stakeholders are, the more sustainable the final project will be for generations to come.

The business world sometimes uses the term 'triple bottom line' for referring to sustainable economic development. Triple bottom line supports the idea that people, profits, and the planet should all benefit from a project or any economic activity. While some environmentalists might downplay the need for profit and emphasize protection of the environment in the present and for future generations, money is essential for building economic security, especially in very impoverished nations and communities. What is also important is that all people profit economically (not just the elite few) and that fair wages, safe working conditions, and other important workers' and ordinary citizens' rights be respected and considered in the process.

More and more businesses in India and around the world are incorporating the concept of triple bottom line into their work and showing that indeed valuing the environment and people can go hand in hand with development. The struggle for the public occurs when leaders of a business or project being proposed are not aware of these concepts and tend to value profit to themselves over the other two areas. At this point, true consensus building and conflict resolution skills are needed, as well as a strategy for educating and trying to open the mind-set of a developer whose foremost concern is profit. This is not an easy task. As green business development and sustainability continue to become more and more mainstream, this will become easier. The government, CII, and many NGOs are promoting sustainable development in India. However, at times it will fall in the hands of citizens at the grassroot level to force their concerns into the minds and hearts of the developers.

7.6.4 Public Participation and Litigation

Examples of public participation and litigation activities provide insight into how public participation with sustainability concerns can proceed. The challenges should not dissuade a student of environmental studies from engaging in such activism. While success is not always a given, constant voicing of this perspective and a desire to consider environmental impacts can lead to reflection and hopefully the sustainable outcomes one desires.

An example of public participation that did not successfully sway the minds of a developer can be found in the community of East Harlem, New York City. The Blumenfeld Development Group purchased the Washburn Wire Factory, at a federal sale in 1996. Built in 1903, the factory, but was then abandoned in 1982. Considered as a brownfield or a former industrial site, it was a decaying dilapidated blight in this impoverished community. There were other plans, including turning it into a movie theatre or movie production centre. The developer planned to build an approximately 470,000 ft^2 retail or what is known as a big box shopping centre with anchor tenants. Other smaller retail shops would be located in the plaza as well. About $40 million of federal, state, and city money was to be spent to build a parking garage for the project and other items as part of the community development funds.

While it was a privately-owned business, the developer did have to get approval for certain elements of the project by the local community board to proceed. Other levels of government also had a vote on whether to approve the requests or not. Some other property owners were slated to lose their land through eminent domain (the process of the government taking private land for public use). Other approvals were needed to get funding for the large parking garage on the site and to modify traffic regulations.

At the community board level, the issue was first addressed in a small taskforce. Members were asked to volunteer to be on the committee to prepare a report and viewpoint on the project to be brought before the entire community board. The group held many meetings, conducted much research, and spoke to many experts. In the end, the committee prepared a report and voted to request that the entire community board vote 'NO' on the project. There were many pros and cons addressed in the report. The following table summarizes these issues.

Pros	Cons
• It will create approximately 1200–2000 jobs, both construction and retail jobs. • The entry-level jobs will be a good starting point for teenagers and low-skilled workers. • Opportunities will exist for many to move to management-level jobs.	• When land is taken to build the plaza, 140 jobs will be lost at businesses that must close. • Other similar businesses may suffer in other parts of Harlem. • Developer and businesses at plaza have not promised to hire a certain percentage of people from East Harlem. • Many retail jobs are typically part time, no benefits, few managers, and little advancement to better paying jobs for the society.
• It will bring more people to East Harlem, which will stimulate business at other stores.	• Will Harlem residents shop at Home Depot and Costco? Few own homes and Costco charges membership, does not accept food stamps, and requires large storage space for bulk items. • Some stores may close when plaza opens.
• It will remove an unsightly, abandoned building.	• Retail atmosphere and traffic will change the feeling of a residential, historical neighbourhood. • The project does not provide access to the waterfront. • It also requires taking of land from private property owners and current businesses, which must close or relocate. 'Eminent Domain'—when state takes the land for public good.
• Taxes from retail shopping and real estate will provide income for the government.	• $40 million of public money will be spent to build a parking garage. • Most profit will go out from the community to the developer who is paid rent for space and businesses that profit from sales—not local residents. • There may be increased costs of repairing roads from traffic, increased emergency room visits with asthma attacks.
• It will bring more variety of shopping to East Harlem residents, as well as other NYC residents.	• Residents need more affordable housing, bookstores, gyms, hotels, and businesses or factories that teach technical skills. Some type of green business might be a healthier alternative. • Home Depot and Costco are more for people who live elsewhere in the city, but East Harlem will suffer the pollution and disruption.

Contd...

Contd...

Pros	Cons
• Streets will be improved and more traffic lights installed as roads are redesigned to accommodate trucks and traffic.	• Providing another building will not offset additional 100+ trucks and 3000 cars expected per day. • Traffic gridlock and more trucks create more accidents, especially since there are 14 schools in the area. • Pollution can be expected and will aggravate asthma. Asthma hospitalizations are the highest in the nation in East Harlem.
• The developer will also build or renovate another building for use by not for profits.	• The method of decision-making did not adequately involve community. As a former brownfield site, it should have been the case. People were told what was coming, not surveyed as to what needs were.
• The developer will comply with all laws to curb air and noise pollution under the State Environmental Quality Review Act.	• Few measures such as planting more trees (to clean air), providing sound-proof glass or air-conditioners (which would allow windows to be closed and block pollution) to schools and residents have been promised in writing. • Follow-up and results of state review have been challenged by many experts in these fields.
• Home Depot often donates money to school groups. Schools can get memberships to Costco and shop in bulk for events and fundraisers.	• Safety and health concerns for children are great. There are 14 schools and day care centres nearby, with approximately 5000 children. • What effect will noise have on residents and especially on students' ability to learn at nearby schools? • Will there be more accidents from trucks? • Will absences from school increase due to asthma attacks and skipping school to hang out and shop at the plaza?

After further debate and discussion, the full community board rejected the findings of the committee and voted (50–5) in favour of the project. Then the project went to the City Planning Commission, then to the City Council, later to the Borough President, and finally to the Mayor. All voted in favour of the project. The project also went through the state environmental review board. Testimony against the project was again heard, but the government found no reason to halt the project. Another historical preservation group worked for several years after this to challenge the right of the developer to destroy this historical building. They promoted saving the architectural features and creating a mixed-use development. While they tied the developer up in court for several years, they were not successful. The project did go forward. The community board committee included many green suggestions in its report, highlighting the fact that more housing was needed, the lack of bookstores, and places for exercise in the community.

If one analyses the chain of events that took place, one could say that in 1998 there were not enough interest or understanding of the importance of sustainable development. The committee's testimony and report were submitted to the head of the company, but these did not sway him. Because he privately owned the land, the decision on what and how to build the project could only be suggested by citizens, not forced by the government. Perhaps if the report was read in more recent times, the developer would have been more open to reconsidering the nature of his project. Had the developer truly built consensus and created something really beneficial to the community that incorporated environmental concerns and historical preservation, the final project could have brought him more fame as a model of urban renewal. It certainly would have been a project the entire community would have been proud of and perhaps a tourist attraction. But such lack of understanding and openness to others concerns on the part of developers is a constant reality and presents the challenges and struggle of public participation.

In India, there are some strong decisions that have come about through public interest litigations. India's Constitution protects the right to life, obligates the state to secure the health of its citizens, and defines the need of the state to protect and improve the environment. This establishes the Indian Constitution as one of only a few in the world that contains specific provisions for environmental protection. One of the most successful uses of this provision was led by M C Mehta, a lawyer and environmental activist who filed a case in 1987 on behalf of citizens and tannery workers concerned about the pollution of River Ganga. The state ordered the tanneries to close if they could not correct and stop releasing polluting effluents. In the ruling, the concerns that employment and economic security would be severely disrupted in the region were deemed less important than the environmental disruption caused by the pollutants and seen as a violation of the Indian Constitution.

The complexities of public participation and involvement in environmental issues were also seen in the case of converting the fuel-powered New Delhi's transit system to CNG (Figure 7.10). Again the environmental lawyer M C Mehta brought the public interest case to the Supreme Court in 1985 as an attempt to improve New Delhi's air pollution, which was said to be claiming the lives of 10,000 citizens each year.

Figure 7.10 CNG filling station
Picture courtesy Lynn Tiede

Public outcry and World Health Organization reports over high levels of air pollution led to filing of the case. Mr Mehta won the case, and the Supreme Court mandated that the government must phase out diesel fuel and have all city buses, motorized rickshaws, and other taxis convert to CNG.

Unfortunately, the government ignored the ruling, allowed the sale of diesel vehicles, and did not build the needed infrastructure to provide CNG. When the conversion deadline came in 2001 and all diesel vehicles were banned, chaos ensued. Citizens could not get to work or school and drivers lost jobs. The government and public outcry pushed for an extension of the deadline, and the court was accused of overstepping its role. Since then, New Delhi has converted to CNG, but air pollution issues still plagues the city and India as the overall numbers of all types of vehicles on the road increases. Constant vigilance and competing interests are always at play but do not diminish the need for ongoing public participation and oversight of environmental issues.

7.7 CONCLUSION

In the chapter, difficulty of building consensus and balancing the needs of people, profit, and the environment in the modern world have been discussed with examples. There will always be a struggle, and always entities that believe a business has not gone far enough or that they have compromised some level of sustainability in one area of the project. However, the more input, the more information, research, and perspectives that are included in the project, the more beneficial the final project can become for all affected. The challenge of public participation in a democracy should not lull us into apathy. These are very hopeful times. Many people who have been active in social justice and environmental movements since the 1970s believe that the consciousness of the world is shifting, despite the difficulties encountered at times. Such terms as the 'green revolution' or 'envirolution' are used by some. The World Social Forum, an annual meeting of small grassroots organizations challenging globalization, aptly coined the phrase "another world is possible". With the input of all, a sustainable economy and lifestyle for all can be built.

8

Climate Change and its Impacts

8.1 INTRODUCTION

Most scientists and politicians agree that climate change is happening. How is this apparent? The average global temperature has increased; seven of the hottest years in the recorded history have been witnessed since 2000. Sea levels are rising as ice is shrinking in Antarctica and Greenland. These are facts based on careful measurements. Do humans notice these changes visibly? Probably not since climate describes long-term shifts in temperature. What is noticeable is the weather in the form of daily and weekly shifts that mask the long-term changes.

Notice that the words "global warming" were not used. Why not if average temperatures across the globe are increasing? The key is the word "average". In some places, temperatures may be below average for a while. The United States had that situation early in 2015 when the eastern parts of the country were cold. Boston had record cold and snowfall, and some residents thought "global warming" was not happening. However, they missed the point that during the same period, western United States experienced record high temperatures and low rainfall. California suffered from a severe drought, and Anchorage (Alaska) at the latitude 58° north was warmer than Chicago at 42°. Extremes are one of the results of climate change. India has also been experiencing dramatic changes with severe drought in many parts of the country.

Since the Kyoto Protocol in 1997, efforts are being made continuously at international forums to find ways to minimize climate change and its impacts on humans and the environment. The targeted increase in average temperature is 2°C, since any temperature above this will result in the melting of ice caps and the flooding of coastal areas in many countries. Carbon dioxide in the atmosphere has dramatically increased since the last century as industrialization progressed in many parts of the world, especially in the western hemisphere (see Figure 8.1). Since India gained independence in 1947 and rapid industrialization followed and China emerged as a major industrial nation in the 1980s, the impacts of climate change have increased.

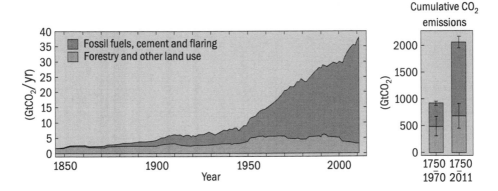

Figure 8.1 Increase in carbon dioxide emissions in the 20th century
Note Quantitative information on CH_4 and N_2O emission time series from 1850 to 1970 is limited

8.2 INTERGOVERNMENTAL PANEL ON CLIMATE CHANGE

National and international scientific societies have assessed current scientific opinion on global warming. The Intergovernmental Panel on Climate Change (IPCC) was established in 1988 by the World Meteorological Organization (WMO) and the United Nations Environment Programme (UNEP) to assess scientific, technical, and socio-economic information concerning climate change, its potential effects, and options for adaptation and mitigation. The assessments of national and international groups were generally consistent with the conclusions of the IPCC. The *First Assessment Report* of the IPCC was published in 1990, while the second, third, and fourth reports were published in 1995, 2001, and 2007, respectively. The *Fourth Assessment Report* (2007) summarized the following:

- Warming of the climate system is unequivocal as evidenced by increases in global average air and ocean temperatures, the widespread melting of snow and ice, and rising global average sea level.
- Most of the global warming since the mid-20th century is very likely due to human activities.
- Benefits and costs of climate change for (human) society will vary widely by location and scale. Some of the effects in temperate and polar regions will be positive, and others elsewhere will be negative. Overall, the net effects are more likely to be strongly negative with larger or more rapid warming.

- The range of published evidence indicates that the net damage costs of climate change are likely to be significant and increase over time.
- The resilience of many ecosystems is likely to be exceeded this century by an unprecedented combination of climate-change-associated disturbances (for example, flooding, drought, wildfire, insects, ocean acidification) and other global change drivers (for example, land-use change, pollution, fragmentation of natural systems, overexploitation of resources).

The IPCC *Fifth Assessment Report* was finalized in 2014. The report is based on 9200 peer-reviewed reports, and its principal findings are as follows:

- Warming of the atmosphere and ocean systems is unequivocal. Many of the associated impacts such as sea-level change (among other metrics) have occurred since 1950 at rates unprecedented in the historical record.
- There is a clear human influence on the climate.
- It is extremely likely that human influence has been the dominant cause of observed warming since 1950, with the level of confidence having increased since the fourth report.

The IPCC pointed out that the longer the countries wait to reduce the emissions, the more expensive it will become. The fifth IPCC report was based on some historical metrics, which are summarized as follows:

- It is likely (with medium confidence) that 1983–2013 was the warmest 30-year period for 1400 years.
- It is virtually certain the upper ocean warmed from 1971 to 2010. This ocean warming accounts, with high confidence, for 90% of the energy accumulation.
- It can be said with high confidence that the Greenland and Antarctic ice sheets have been losing mass in the last two decades, and the Arctic ice sheets and Northern Hemisphere spring snow cover have continued to decrease in extent.
- The sea-level rise since the middle of the 19th century has been larger than the mean sea-level rise of the prior two millennia.
- Concentration of greenhouse gases (GHGs) in the atmosphere has increased to levels unprecedented on earth in 800,000 years.
- The total radiative forcing of the earth system, relative to 1750, is positive, and the most significant driver is the increase in atmospheric concentration of carbon dioxide.

While historical evidence regarding climate change is large as seen from the fifth IPCC report, much less information is available about the extent of the effects: how much will happen, when? Here is one that could be serious for India: seasonal flow in some rivers. The River Ganga is of concern since some of its seasonal flow comes from glaciers high in the Himalayas. If these glaciers shrink to a small fraction of their current size, river flow will also fluctuate wildly—high during the monsoons and from melt snow in the mountains. When those cease each year, there would be little glacial melt to maintain the flow. While the glaciers are shrinking, there is uncertainty about how long they will last. Decades? Centuries?

8.3　EFFECTS OF CLIMATE CHANGE ON HUMANS

Climate change has resulted in large-scale environmental hazards to human health, such as extreme weather, ozone depletion, increased danger of wildfires, loss of biodiversity, stresses to food-producing systems, and the global spread of infectious diseases. The World Health Organization estimates that 160,000 deaths since 1950 are directly attributable to climate change. Various groups around the world are quantifying the impacts of climate change on health, food supply, economic growth, migration, security, societal change, and public goods such as drinking water. However, it is certain that the majority of the adverse impacts will be experienced by the poor and low-income communities around the world since they have very less capacity for coping with environmental change. The Global Humanitarian Forum in Geneva published a report in 2009 that estimated more than 300,000 deaths and about $125 billion in economic losses each year due to worsening floods and droughts in developing countries.

Could other unusual effects happen in India? Possibly. Maybe climate change will make the monsoons weaker or stronger, or shut them off. When the monsoon fails, crops fail and India suffers episodes of starvation. Rising sea levels will cover coastal areas, including the Odisha coast and the Sunderbans at the mouth of Ganges, shared with Bangladesh. Storm surges and drenching rains have flooded large parts of Bangladesh, which are already not much above sea level; when sea levels rise, the ocean moves in and storm surges devastate substantial areas. Will this affect India directly? Bangladesh is already crowded; if citizens are forced off their land, one place they will flee is to India, especially Kolkata and in other parts of West Bengal.

What is behind the consensus that climate change is upon us? What is the cause? Simply put, carbon dioxide is causing greenhouse effect. Like the glass roof of a greenhouse, carbon dioxide is transparent to sun rays. The energy of those rays striking the earth turns to heat, and carbon dioxide blocks the return of the heat to space, just as the glass roof in a greenhouse holds the warmth. One of the major sources of carbon dioxide is burning of fossil fuels. Carbon dioxide is emitted when petroleum is burnt by vehicles. Production of electricity is another source. When coal burns, thermal energy is generated. Of course, the exhaust from burning coal is carbon dioxide.

How much carbon dioxide is present in the atmosphere? It is only about 0.004%, compared to oxygen at about 20%. Usually the amount is given in parts per million (ppm) with the current value around 400 ppm. Carbon dioxide concentration has been tracked for over five decades, starting in 1958 by Charles Keeling when he measured the value of 315 ppm. The resulting graph, called the Keeling Curve, shows the rise in carbon dioxide over time (Figure 8.2). The projection is for more increase.

8.4 PREVENTION AND MITIGATION OF CLIMATE CHANGE IMPACTS

As noted earlier, climate change has been severely impacting humans, animals, and the environment. What can be done? International cooperation is essential since carbon dioxide from any country spreads

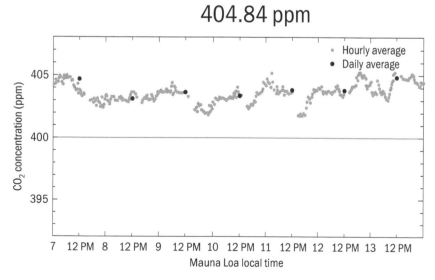

Figure 8.2 Keeling Curve showing carbon dioxide concentration at Mauna Loa observatory
Note The reading is till 13 April 2015
Source <https://en.wikipedia.org/w/index.php?title=Keeling_Curve&oldid=677032704>

across the globe. A key conference was the Earth Summit in Rio de Janeiro in 1992, which established the United Nations Framework Convention on Climate Change, followed by the Kyoto Protocol in 1997, which set some specific guidelines for GHGs. And what are the solutions? Some of them include reduced use of fossil fuels by substituting solar, wind, and nuclear power; reduced per capita energy use; improved public transportation; and lock up of carbon dioxide by planting more trees and grasslands. All have been suggested and perhaps all will be needed. In the meantime, carbon dioxide levels continue to rise on planet earth.

Reducing the use of fossil fuels has gained a lot of attention worldwide, and a huge push is witnessed for the use of solar technologies, not only for individual homes but also for large-scale power production. Several new technologies have been developed, some with subsidies by governments and some by entrepreneurs seeking new opportunities in alternative energy development. A large solar collector system, coupled with power generation, is in operation in the western United States. The National Renewable Energy Laboratory (NREL) in Boulder, Colorado, has been developing new concepts for increasing the use of solar power. Its projections for shared solar projects, where a community-installed solar project provides solar energy to households and businesses in the community, are shown in Figure 8.3.

The community of Brewster, Massachusetts, has installed a solar garden to supply energy to households in the community (Figure 8.4). Several power utilities in the United States have been using hybrid technologies, using solar power along with clean burning natural gas. The cost of solar panels has reduced as demand has increased in China, India,

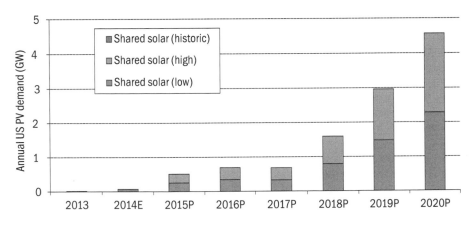

Figure 8.3 NREL projections for shared solar projects
Source Feldman, Brockway, Urlich, *et al.* (2015)

Figure 8.4 Brewster Community Solar Garden
Source <www.brewstercommunitysolargarden.com/from-concept-to-reality>

Europe, and the United States. Rural areas of countries rely on solar power only since they are not connected to the electrical grid. Several solar technologies have been developed to provide for heating and cooling of homes and buildings (which uses a lot of energy generated from fossil fuels) and for cooking food. Governments have been providing subsidies for promoting solar power, with the Indian Government providing as much as 30% subsidy for solar power use.

Wind energy has grown in popularity all around the world, with wind power being popular in Europe, China, the United States, and India. A number of coastal wind energy projects have been planned. Land-based wind energy towers have grown in number around the world and are producing substantial amount of energy. China is using hybrid wind and solar energy for street lighting, thus saving a lot of fossil-generated energy. Subsidies for wind energy have been provided by many governments.

Nuclear energy has become safer, and smaller power plants using nuclear energy have been developed. India has stepped up with more nuclear energy plants with support from the international community. Nuclear energy can replace many large fossil-fuelled power plants in India and China. Reuse and disposal of nuclear waste still pose a huge problem, and many governments have been working on solutions with France taking the leading role.

Energy efficiency has received a lot of attention in the United States, Europe, China, and India since every watt saved through efficiency is one watt that has to be generated using fossil fuel. Incandescent and fluorescent bulbs are being replaced with light-emitting diodes in commercial and residential buildings, and this saves almost two-thirds of the current energy use. Other ways of saving energy include use of sensors to turn off lights when no one is present in a room and improved heating and cooling systems for buildings.

Green electric transportation and fuel efficiency in diesel and gasoline vehicles have been promoted worldwide. Governments play a major role in setting fuel efficiency standards for personal and commercial vehicles. This has saved a lot of fossil fuels. In addition, the use of natural gas has increased, thus reducing GHG emissions. Electric vehicles are still expensive, and as battery technologies have improved and costs have further reduced, more electric transportation will replace diesel or gasoline vehicles.

Carbon sequestration has been attempted to capture the carbon dioxide emitted from power plants using fossil fuels. The United States is evaluating the economics of carbon sequestration; however, the technology is still expensive and not being adopted by power plants. Many governments have pursued cap-and-trade systems for reducing overall GHGs in their areas. Highly polluting industries that cannot economically retrofit their facilities pay money to more efficient industries so that the overall level of GHG emissions is kept under a cap established by the government. However, market forces play a big part in the success of cap-and-trade systems.

Under global agreements, reforestation projects have been undertaken in many parts of the world since forests act as GHG sinks. Incentives for using biogas from animal and human wastes are also provided under these agreements for reducing the use of wood as a cooking fuel. Improved agricultural practices have also ceased large-scale deforestation in many parts of the world. Vertical farms have been developed in abandoned industrial buildings for growing food locally and eliminating the need for transportation over long distances. Prevention of deforestation combined with reforestation efforts has positively influenced the reduction in GHGs.

A global agreement to reduce greenhouse gases by 30% till 2050 was signed in December 2015. This will force governments to take all steps necessary, including economic incentives, to provide alternative energy sources and efficient energy use to reduce GHG emissions. It is encouraging that China and the United States signed an agreement in early 2015 to work towards the goal of reducing GHG emissions. India has taken an active part in the global negotiations, and it is hoped that this will result in the world's largest emitters of GHGs to drastically reduce their emissions. Europe, China, India, and the United States are working together to tackle this global problem, and it is a very encouraging sign for the future.

Many countries are spending large sums of money to mitigate the rise in sea level. The Italian Government is building a large wall to keep the rising sea level from adversely impacting the city of Venice. The Netherlands Government is building dikes to ensure that rising sea levels do not adversely impact its cities. A recent public interest litigation (PIL) filed against the Netherlands Government forced the government to take more proactive prevention steps to reduce GHGs. PILs have been used effectively to improve the air and water quality in India. As the citizenry becomes more alarmed with the adverse impacts of climate change, governments would be forced to take actions to reduce GHGs. Local and state governments all over the United States have taken preventive steps and set aggressive goals to improve energy efficiency and reduce GHG emissions. This is a very encouraging sign that citizens and governments at all levels are participating and cooperating to reduce the emission of GHGs. Since poor countries cannot mitigate the impact of rising sea levels and other extreme weather conditions resulting from climate change, richer nations should share the knowledge gained by research with the poor countries and help them economically in reducing GHGs. All inhabitants living on this planet play a part in preventing the increase in GHGs, both individually and as communities. This is imperative that a habitable planet continues to exist for future generations. People's active involvement as global citizens is required for preserving this planet for future generations.

9

Environmental Legislation and Enforcement

9.1 INTRODUCTION

The Constitution of India enshrines under:

Article 48A: The State shall endeavour to protect and improve the environment and safeguard the forests and wildlife of the country; and

Article 51A(g): Fundamental duty of every citizen to protect and improve the natural environment, including forests, lakes, rivers, and wildlife and to have compassion for living creatures.

9.2 LAWS FOR ENVIRONMENTAL MANAGEMENT

In India, the Ministry of Environment, Forests, and Climate Change (MoEFCC) is a multidisciplinary regulatory body for environmental protection and conservation in the country. Environmental laws of unprecedented scope and impact have been passed by the Government of India to protect and improve the environment. The relevant legislations for environmental management are summarized as follows:

9.2.1 Water (Prevention and Control of Pollution) Act, 1974

Rapid depletion and degradation of water quality and its ill effects on human health forced the authorities to introduce the first legislation to prevent and control water pollution in as early as the mid-1970s. The Water (Prevention and Control of Pollution) Act, 1974 provides for the prevention and control of water pollution and the maintenance or restoration of wholesomeness of water. As such, all human activities having a bearing on water quality are covered under this act.

Subject to the provisions of the act, no person, without the prior consent of the state pollution control board (SPCB), can establish any industry, operation, or process, or any treatment and disposal system or an extension or addition thereto, which is likely to discharge sewage or trade effluent into a stream or well or sewer or on land, and one has to apply to the SPCB concerned to obtain the 'Consent to establish' as well as the 'Consent to operate' the industry after establishment.

9.2.2   Water (Prevention and Control of Pollution) Cess Act, 1977

The main purpose of the Water (Prevention and Control of Pollution) Cess Act, 1977 is to levy and collect cess on the water consumed by every person/local authority. The money thus collected is used by the SPCBs to prevent and control water pollution.

9.2.3   Water (Prevention and Control of Pollution) Cess (Amendment) Act, 2003

In the amendment, 'industry' includes any operation or process or treatment and disposal system that consumes water or gives rise to sewage effluent or trade effluent but does not include any hydropower unit.

In the Principal Act, the word 'industry' is substituted for 'specified industry'. For Section 16 of the Principal Act, the following section is substituted:

Power of Central Government to Exempt the Levy of Water Cess:

1. "16.(1) Notwithstanding anything contained in Section 3, the Central Government may, by notification in the Official Gazette, exempt any industry, consuming water below the quantity specified in the notification, from the levy of water cess.

2. "Schedule I to the Principal Act is omitted and Substitution of New Schedule for Schedule II."

9.2.4   Air (Prevention and Control of Pollution) Act, 1981

The objective of the Air (Prevention and Control of Pollution) Act, 1981 is to prevent, control, and reduce air pollution, including noise pollution. Under the provisions of this act, no person shall, without the previous consent of the SPCBs, establish or operate any industrial plant in an area under air pollution control. The investor has to apply to the SPCBs/Pollution Control Committee (PCC) to obtain consent. No person operating any industrial plant shall emit any air pollutant in excess of the standards laid down by the SPCBs; the person has to comply with the stipulated conditions.

9.2.5   Environment (Protection) Act, 1986

The Environment (Protection) Act, 1986 is an umbrella act for the protection and improvement of environment and for matters connected with it. It provides that no person carrying on any industry, operation, or process should discharge or emit or should be permitted to discharge or emit any environmental pollutant in excess of prescribed standards.

Several sets of rules relating to various aspects of management of hazardous chemicals and wastes have also been notified. The standards for pollutants are to be achieved within a period of 1 year from the date of their notification, especially by those industries identified as highly polluting. However, if a particular SPCB desires, it may reduce the time limit and also specify more stringent standards for a specified category of industries within its jurisdiction. The SPCB, however, cannot relax either the time limit or the standards stipulated by the Government of India. Section 15 ensures punishment, fine, and imprisonment for the violation of the act.

Subject to the provision of this act, the Central Government has the power to take all measures as it deems necessary or expedient for the purpose of protecting and improving the environment and preventing, controlling, and abating environmental pollution.

Procedures, safeguards, prohibition, and restriction on the handling of hazardous substances along with the prohibition and restriction on the location of industries and carrying on processes and operations in different areas have been notified. Restrictions have been imposed on various activities in fragile areas, such as Doon Valley in Uttarakhand; Aravali Regions in Alwar, Rajasthan; coastal zones; and ecologically sensitive zones. Besides, the Public Liability Insurance Act, 1991 is constituted to provide immediate relief to the persons affected by accidents occurring while handing any hazardous substance.

9.2.6 Hazardous Wastes (Management and Handling) Rules, 1989

The Hazardous Wastes (Management and Handling) Rules, 1989 came to be known as the Hazardous Wastes (Management, Handling and Transboundry Movement) Rules, 2008 from 24 September 2008.

Various forms (given below) are used for documentation:
- Form 1 for obtaining authorization for collection/treatment/storage/disposal/transport of hazardous waste,
- Form 2 for grant/renewal of authorization,
- Form 3 for maintaining proper records,
- Form 4 for filing annual returns,
- Form 5 for grant/renewal of registration of industrial units,
- Form 6 for filing annual returns of annual records on recyclable hazardous waste,
- Form 7 for import or export of hazardous waste,
- Form 8 for the transboundry movement of hazardous waste,

- Form 9 for the transboundry movement of documents,
- Form 10 for maintaining records of hazardous waste imported and exported,
- Form 11 for Transport Emergency Card,
- Form 12 for marking of hazardous waste container,
- Form 13 for manifesting hazardous wastes,
- Form 14 for reporting accidents,
- Form 15 for filing appeal against the order passed by the Central Pollution Control Board (CPCB)/SPCB/PCC, and
- Form 16 for registration of traders.

The MoEFCC notified the Hazardous Wastes (Management and Handling) Amendment Rules on 6 January 2000. Under these rules, toxic chemicals, flammable chemicals, and explosives have been redefined to be termed as 'hazardous chemical'. As per the new criteria, 684 hazardous chemicals have been identified instead of 4343 chemicals listed in the Hazardous Wastes Rules, 1989. The hazardous substances have been put in three categories: (i) process-specific industrial wastes, (ii) waste substances with concentration limits, and (iii) wastes applicable only for imports and exports. The authorization application shall be processed by the SPCBs within 90 days. It will be valid for 5 years, and its renewal will depend on steps taken for reduction in the waste generated, recycled, or reused. Disposal sites for hazardous wastes shall be identified by the State Government (operator) of a facility or occupier. An environmental impact assessment (EIA) is to be carried out for selecting the appropriate site. Public hearing for objections and suggestions has to be arranged by the SPCBs within 30 days. The SPCBs will monitor the setting up and operation of a facility regularly. The operation and closure of landfill site should be carried out by the SPCBs as per Rule 8A. The import and export of hazardous wastes for dumping and disposal are strictly prohibited. These are permitted only if the raw materials are used for recycling or reuse.

9.2.7 Manufacture, Storage, and Import of Hazardous Chemicals Rules, 1989 and 2000

Under the Manufacture, Storage, and Import of Hazardous Chemical Rules, 1989 and 2000, project proponents of any kind of hazardous industry have to identify likely hazards and their danger potential. They also have to take adequate steps to prevent and limit the consequences of any accident at the site. Information regarding accidents is to be updated as per Schedule 7. Material safety data sheets for all the chemicals handled have to be prepared. Onsite workers are required to be provided with information, training, and necessary equipment to ensure their safety. An onsite emergency plan is to be prepared before initiating any activity at the site. An offsite emergency plan is to be prepared by the district collector in close collaboration with the project proponents for any accident envisaged onsite. The public in the vicinity of the plant should be informed of the nature of major accidents that may occur onsite and the 'Dos and Donts' to be followed in the case of such an occurrence. The import of hazardous chemicals is to be reported to the concerned authority within 30 days from the date of import.

The MoEFCC made significant amendments to the Manufacture, Storage, and Import of Hazardous Chemicals Rules, 1989 on 20 January 2000. Under the new amendments, a new Schedule I has been incorporated as a result of the increase in the number of hazardous chemicals. Renewal of authorization will be subject to submission of 'Annual Returns' for disposal of hazardous wastes; production of evidence of reduction in the waste generated or recycled or reused; fulfilment of authorization conditions; and remittance of processing and analysis fee. The State Government as well as the occupier or its association shall be responsible for the identification of site for common waste disposal facility. Public hearing by the State Government is also made mandatory before notifying any common hazardous waste disposal site as per the procedure laid down in Gazette Notification dated 10 April 1997. The central/state government will provide guidance for the design, operation, and closure of common waste facility/landfill site. It is mandatory to obtain prior approval from the SPCB for the design and layout of the proposed hazardous waste disposal facility. A comprehensive procedure has also been laid down in the Manufacture, Storage, and Import of Hazardous Chemicals Rules, 2000 for the regulation of export and import of hazardous wastes.

9.2.8 Public Liability Insurance Act, 1991

The Public Liability Insurance Act, 1991, unique to India, imposes on the owner the liability to provide immediate relief in respect of death or injury to any person or damage to any property resulting from an accident while handling any of the notified hazardous chemicals. This relief has to be provided on 'no fault' basis. The owner handling hazardous chemicals has to take an insurance policy to meet this liability of an amount equal to its 'paid up capital' or up to ₹500 million, whichever is less. The policy has to be renewed every year. A new undertaking has to take this policy before starting its activity. The owner also has to pay an amount equal to its annual premium to the Central Government's Environment

Relief Fund (ERF). The reimbursement of relief to the extent of ₹25,000 per person is admissible in the case of fatal accidents in addition to the reimbursement of medical expenses up to ₹12,500 per person. The liability of the insurance is limited to ₹50 million per accident up to ₹150 million per year or up to the tenure of the policy. Any claims in excess to this liability will be paid from the ERF. In case the award still exceeds, the remaining amount has to be met by the owner. The payment under the act is only immediate relief; owners have to provide the final compensation, if any, arising out of legal proceedings.

9.2.9 Prior Environmental Clearance, Notification dated 14 September 2006

S.O. 1533: A draft notification under Sub-rule (3) of Rule 5 of the Environment (Protection) Rules, 1986 for imposing certain restrictions and prohibitions on new projects or activities, or on the expansion or modernization of existing projects or activities based on their potential environmental impacts as indicated in the schedule to the notification, being undertaken in any part of India,[1] unless prior environmental clearance has been accorded in accordance with the objectives of National Environment Policy as approved by the Union Cabinet on 18 May 2006 and the procedure specified in the notification, by the Central Government or the state- or union territory-level environment impact assessment authority, to be constituted by the Central Government in consultation with the State Government or the union territory administration concerned under Sub-section (3) of Section 3 of the Environment (Protection) Act, 1986 for the purpose of this notification, was published in the Gazette of India, Extraordinary, Part II, Section 3, Sub-section (ii) vide number S.O. 1324 (E) dated 15 September 2005 inviting objections and suggestions from all persons likely to be affected thereby within a period of 60 days from the date on which copies of Gazette containing the said notification were made available to the public.

Copies of the said notification were made available to the public on 15 September 2005.

And all objections and suggestions received in response to the above-mentioned draft notification have been duly considered by the Central Government.

Now, therefore, in exercise of the powers conferred by Sub-section (1) and Clause (v) of Sub-section (2) of Section 3 of the Environment (Protection) Act, 1986, read with Clause (d) of Sub-rule (3) of Rule 5 of the Environment (Protection) Rules, 1986 and in supersession of the

---

[1] Includes the territorial waters

notification number S.O. 60 (E) dated the 27 January 1994, except in respect of things done or omitted to be done before such supersession, the Central Government hereby directs that on and from the date of its publication, the required construction of new projects or activities or the expansion or modernization of existing projects or activities listed in the schedule to this notification entailing capacity addition with change in process and or technology shall be undertaken in any part of India only after the prior environmental clearance from the central government or as the case may be, by the state environment impact assessment authority (SEIAA), duly constituted by the Central Government under Sub-section (3) of Section 3 of the act, in accordance with the procedure specified hereinafter in this notification.

9.2.9.1 Requirement of prior environmental clearance

Projects or activities shall require prior environmental clearance from the concerned regulatory authority, which is the Central Government in the MoEFCC for matters falling under Category A in the schedule and SEIAA for matters falling under Category B in the said schedule, before any construction work or preparation of land by the project management, except for securing the land, is started on the project or activity.

9.2.9.2 State environment impact assessment authority

An SEIAA shall be constituted by the Central Government under Sub-section (3) of Section 3 of the Environment (Protection) Act, 1986, comprising three members, including a chairman and a member–secretary to be nominated by the state government or the union territory administration.

- The member secretary shall be a serving officer of the concerned state government or union territory administration familiar with environmental laws.
- The chairman shall be an expert in terms of the eligibility criteria given in Appendix VI[2] in one of the specified fields, with sufficient experience in environmental policy or management.
- The other members shall be an expert fulfilling the eligibility criteria given in Appendix VI to this notification.
- The state government or union territory administration shall forward the names of the members and the chairman to the Central Government, and the Central Government shall constitute the SEIAA as an authority for the purpose of this notification within 30 days of the date of receipt of the names.

---

[2] Ministry of Environment and Forests. 2006. Published in the Gazette of India, Extraordinary, Part-II, and Section 3, Sub-section (ii) New Delhi: Ministry of Environment and Forests

- The non-official member and the chairman shall have a fixed term of 3 years (from the date of publication of the notification by the Central Government constituting the authority).
- All decisions of the SEIAA shall be taken in a meeting and ordinarily be unanimous provided that, in case a decision is taken by majority, the details of views, for and against it, shall be clearly recorded in the minutes and a copy thereof sent to the MoEFCC.

9.2.9.3 *Categorization of projects and activities*

- All projects and activities are broadly categorized into two categories, Category A and Category B, based on the spatial extent of potential impacts and potential impacts on human health and natural and human-made resources.
- All projects or activities included in Category A in the schedule, including expansion and modernization of existing projects or activities and change in product mix, shall require prior environmental clearance from the Central Government in the MoEFCC on the recommendations of Expert Appraisal Committee (EAC) to be constituted by the Central Government for the purposes of this notification.
- All projects or activities included in Category B in the schedule, including expansion and modernization of existing projects or activities or change in product mix, but excluding those that fulfil the general conditions stipulated in the schedule, will require prior environmental clearance from the SEIAA, which will base its decision on the recommendations of a state expert appraisal committee (SEAC) to be constituted for in this notification. [In the absence of a duly constituted SEIAA or SEAC, a Category B project shall be treated as a Category A project.][3]

9.2.9.4 *Screening, scoping, and appraisal committees*

The EACs at the Central Government level and SEACs at the state or the union territory level shall screen, scope, and appraise projects or activities in Category A and Category B, respectively. The EACs and SEACs shall meet at least once every month.

- The composition of the EAC shall be as given in Appendix VI. The SEAC at the state or the union territory level shall be constituted by the Central Government in consultation with the concerned state

---

[3] Ministry of Environment and Forests. 2009. The Gazette of India: Extraordinary, notification dated 1 December 2009. New Delhi: Ministry of Environment and Forests

government or the union territory administration with identical composition.

- The Central Government may, with the prior concurrence of the concerned state governments or union territory administrations, constitute one SEAC for more than one state or union territory for reasons of administrative convenience and cost.

- The EACs and SEACs shall be reconstituted after every 3 years.

- The authorized members of the EACs and SEACs concerned may inspect any sites connected with the project or activity for which prior environmental clearance is sought, for the purposes of screening or scoping or appraisal, with prior notice of at least 7 days to the applicant, who shall provide necessary facilities for the inspection.

- The EACs and SEACs shall function on the principle of collective responsibility. The chairperson shall endeavour to reach a consensus in each case, and if consensus cannot be reached, the view of the majority shall prevail.

9.2.9.5 Application for prior environmental clearance

An application seeking prior environmental clearance in all cases shall be made by the project proponent[4] in the prescribed Form 1 and supplementary Form 1A, if applicable, as given in Appendix II[5], after the identification of prospective sites for the project and/or activities to which the application relates, before commencing any construction activity or preparation of land at the site by the applicant. Along with the application, the applicant shall furnish a copy of the pre-feasibility project report except that, in the case of construction projects or activities (item 8 of the schedule) in addition to Form 1 and supplementary Form 1A, a copy of the conceptual plan shall be provided, instead of the pre-feasibility report.

9.2.9.6 Stages in the prior environmental clearance process for new projects

The environmental clearance process for new projects will consist of a maximum of four stages, all of which may not apply to particular cases as set forth in this notification. These four stages in sequential order are as follows:

---

[4] Ministry of Environment and Forests. 2011. Compendium of Gazette Notifications, Office Memoranda under Environment Impact Assessment Notification, 2006, The Gazette of India: Extraordinary, notification dated 4 April 2011

[5] Ministry of Environment and Forests. 2016. Published in the Gazette of India, Extraordinary, Part-II, and Section 3, Sub-section (ii) New Delhi: Ministry of Environment and Forests

- Stage 1: Screening (Only for Category B projects and activities)
- Stage 2: Scoping
- Stage 3: Public consultation
- Stage 4: Appraisal

Stage 1   Screening

In the case of Category B projects or activities, this stage will entail the scrutiny of an application seeking prior environmental clearance made in determining whether or not the project or activity requires further environmental studies for preparation of an EIA for its appraisal prior to the grant of environmental clearance depending on the nature and location specificity of the project. The projects requiring an EIA report shall be termed Category B1 and the remaining projects shall be termed Category B2 and will not require an EIA report. For the categorization of projects into B1 or B2, except item 8(b), the MoEFCC shall issue appropriate guidelines from time to time.

Stage 2   Scoping

- Scoping refers to the process by which the EAC in the case of Category A projects or activities and the SEACs in the case of Category B1 projects or activities, including applications for expansion and/or modernization and/or change in product mix of existing projects or activities, issuance of terms of reference (TOR) as per standard TOR issued by the MoEFCC addressing all relevant environmental concerns for the preparation of an EIA report of the project or activity for which prior environmental clearance is sought. The EAC or SEACs concerned shall determine the TOR on the basis of the information furnished in the prescribed application Form 1/Form 1A, including the TOR proposed by the applicant, a site visit by a sub-group of EAC or SEACs concerned only if considered necessary by the EAC or SEACs concerned, TOR suggested by the applicant, if furnished, and other information that may be available with the EAC or SEACs concerned. All projects and activities listed under Category B in item 8(a) of the schedule (building and construction projects) shall not require scoping and will be appraised on the basis of Form 1/Form 1A and the conceptual plan.
- The standard TOR shall be conveyed to the applicant by the EAC or SEACs concerned within 30 days of the receipt of Form 1/Form 1A and conceptual plan. In the case of Category A hydroelectric projects item 1(c)(i) of the schedule, the TOR shall be conveyed along with the clearance for pre-construction activities. The

approved TOR shall be displayed on the websites of the MoEFCC and the concerned SEIAA.

- Applications for prior environmental clearance may be rejected by the regulatory authority concerned on the recommendation of the EAC or SEACs concerned at this stage itself. In the case of rejection, the decision together with reasons for the rejection shall be communicated to the applicant in writing within 60 days of the receipt of the application.

Stage 3 Public Consultation

- Public consultation refers to the process by which the concerns of affected local people and others who have plausible stake in the environmental impacts of the project or activity are ascertained with the view of taking into account all the materials involved in the project or activity design as appropriate. All Category A and Category B1 projects or activities shall undertake public consultation, except the following:
  - Modernization of irrigation projects [item 1(c)(ii) of the schedule].
  - All projects or activities located within industrial estates or parks [item 7(c) of the schedule] approved by the concerned authorities and which are not disallowed in such approvals.
  - Expansion of roads and highways [item 7(f) of the schedule], which does not involve any further acquisition of land (maintenance dredging provided the dredged material shall be disposed within part limits).
  - All building/construction projects/area development projects (which do not contain any Category A projects and activities) and townships [item 8(a) and 8(b) in the schedule to the notification].
  - All Category B2 projects and activities.
  - All projects or activities concerning national defence and security or involving other strategic considerations as determined by the Central Government.
- The public consultation shall ordinarily have two components:
  (i) A public hearing at the site or in its close proximity (district wise), to be carried out in the manner prescribed in Appendix IV[6], for ascertaining concerns of affected local people.

---

[6] Ministry of Environment and Forests. 2016. Published in the Gazette of India, Extraordinary, Part-II, and Section 3, Sub-section (ii) New Delhi: Ministry of Environment and Forests

    (ii) Obtain responses in writing from other concerned persons having a plausible stake in the environmental aspects of the project or activity.

- The public hearing at or in close proximity to the site(s) in all cases shall be conducted by the SPCB or the union territory pollution control committee (UTPCC) concerned in the specified manner and forward the proceedings to the regulatory authority concerned within 45 days of a request to the effect from the applicant.

- In case the SPCB or the UTPCC concerned does not undertake and complete the public hearing within the specified period and/or does not convey the proceedings of the public hearing within the prescribed period directly to the regulatory authority concerned as above, the regulatory authority shall engage another public agency or authority, which is not subordinate to the regulatory authority, to complete the process within a further period of 45 days.

- If the public agency or authority nominated reports to the regulatory authority concerned that owing to the local situation, it is not possible to conduct the public hearing in a manner that will enable the views of the concerned local people to be freely expressed, it shall report the facts in detail to the concerned regulatory authority, which may, after due consideration of the report and other reliable information that it may have, decide that the public consultation in the case need not include public hearing.

- For obtaining responses in writing from other concerned persons having a plausible stake in the environmental aspects of the project or activity, the concerned regulatory authority and the SPCB or UTPCC shall invite responses from such concerned persons by placing on their website the summary of the EIA report prepared in the format given in Appendix IIIA[7] by the applicant along with a copy of the application in the prescribed form, within 7 days of the receipt of a written request for arranging the public hearing. Confidential information, including non-disclosable or legally privileged information involving intellectual property right, source specified in the application shall not be placed on the website. The concerned regulatory authority may also use other appropriate media for ensuring wide publicity about the project or activity. The regulatory authority shall, however, make available on a written request from any concerned person the draft EIA report

---

[7] Ministry of Environment and Forests. 2016. Published in the Gazette of India, Extraordinary, Part-II, and Section 3, Sub-section (ii) New Delhi: Ministry of Environment and Forests

for inspection at a notified place during normal office hours till the date of the public hearing. All the responses received as part of this public consultation process shall be forwarded to the applicant through the quickest available means.

- After completion of the public consultation, the applicant shall address all the material environmental concerns expressed during this process and make appropriate changes in the draft EIA and environment management plan (EMP). The final EIA report so prepared shall be submitted by the applicant to the concerned regulatory authority for appraisal. The applicant may alternatively submit a supplementary report to draft EIA and EMP addressing all the concerns expressed during the public consultation.

Stage 4   Appraisal

- Appraisal means detailed scrutiny by the EAC or SEACs of the application and other documents such as the final EIA report, outcome of the public consultations, including public hearing proceedings, submitted by the applicant to the regulatory authority concerned for environmental clearance. This appraisal shall be made by the EAC or SEACs concerned in a transparent manner in a proceeding to which the applicant shall be invited for furnishing necessary clarifications in person or through an authorized representative. On conclusion of this proceeding, the EAC or SEACs concerned shall make categorical recommendations to the regulatory authority concerned either for grant of prior environmental clearance on stipulated terms and conditions, or rejection of the application for prior environmental clearance, together with reasons for the decision.

- The appraisal of all projects or activities that are not required to undergo public consultation, or submit an EIA report, shall be carried out on the basis of the prescribed application Form 1 and Form 1A as applicable, any other relevant validated information available and site visit wherever necessary by the EAC or SEACs concerned.

- The appraisal of an application shall be completed by the EAC or SEACs concerned within 60 days of the receipt of the final EIA report and other documents or the receipt of Form 1 and Form 1 A, where public consultation is not necessary and the recommendations of the EAC or SEACs shall be placed before the competent authority for a final decision within the next 15 days. The prescribed procedure for appraisal is given in Appendix V.

All applications seeking prior environmental clearance for expansion with increase in the production capacity beyond the capacity for which prior environmental clearance has been granted under this notification or with increase in either lease area or production capacity in the case of mining projects or for the modernization of an existing unit with increase in the total production capacity beyond the threshold limit prescribed in the schedule to this notification through change in process and/or technology or involving a change in the product mix shall be made in Form 1 and they shall be considered by the concerned EAC or SEACs within 60 days, who will decide on the due diligence necessary, including preparation of EIA and public consultations and the application shall be appraised accordingly for grant of environmental clearance.

9.2.9.7 Grant or rejection of prior environmental clearance

The regulatory authority shall consider the recommendations of the EAC or SEACs concerned and convey its decision to the applicant within 45 days of the receipt of the recommendations of the EAC or SEACs concerned or in other words within 105 days of the receipt of the final EIA report, and where EIA is not required, within 105 days of the receipt of the complete application with requisite documents, except as provided below:

- The regulatory authority shall normally accept the recommendations of the EAC or SEACs concerned. In cases where it disagrees with the recommendations of the EAC or SEACs concerned, the regulatory authority shall request reconsideration by the EAC or SEACs concerned within 45 days of the receipt of the recommendations of the EAC or SEACs concerned while stating the reasons for the disagreement. An intimation of this decision shall be simultaneously conveyed to the applicant. The EAC or SEACs concerned, in turn, shall consider the observations of the regulatory authority and furnish its views on the same within a further period of 60 days. The decision of the regulatory authority after considering the views of the EAC or SEACs concerned shall be final and conveyed to the applicant by the regulatory authority concerned within the next 30 days.

- In the event that the decision of the regulatory authority is not communicated to the applicant within the specified period, as applicable, the applicant may proceed as if the environment clearance sought for has been granted or denied by the regulatory authority in terms of the final recommendations of the EAC or SEACs concerned.

- On the expiry of the period specified for decision by the regulatory authority, as applicable, the decision of the regulatory authority and the final recommendations of the EAC or SEACs concerned shall be public documents.

- Clearances from other regulatory bodies or authorities shall not be required prior to the receipt of applications for prior environmental clearance of projects or activities, or screening, or scoping, or appraisal, or decision by the regulatory authority concerned, unless any of these is sequentially dependent on such clearance either due to a requirement of law or for necessary technical reasons.

- Deliberate concealment and/or submission of false or misleading information or data that are materials for screening or scoping or appraisal or decision on the application shall make the application liable for rejection and cancellation of prior environmental clearance granted on that basis. Rejection of an application or cancellation of a prior environmental clearance already granted, on such ground, shall be decided by the regulatory authority, after giving a personal hearing to the applicant, and following the principles of natural justice.

9.2.9.8 Validity of environmental clearance

"Validity of environmental clearance" means the period from which a prior environmental clearance is granted by the regulatory authority, or may be presumed by the applicant to have been granted, to the start of production operations by the project or activity or completion of all construction operations in the case of construction projects (item 8 of the schedule), to which the application for prior environmental clearance refers. The prior environmental clearance granted for a project or activity shall be valid for a period of 10 years in the case of river valley projects [item 1(c) of the schedule], project life as estimated by the EAC or SEACs subject to a maximum of 30 years for mining projects and 5 years in the case of all other projects and activities. However, in the case of area development projects and townships [item 8(b)], the validity period shall be limited only to such activities as may be the responsibility of the applicant as a developer. This period of validity may be extended by the regulatory authority concerned by a maximum period of 5 years provided an application is made to the regulatory authority by the applicant within the validity period, together with an updated Form 1, and supplementary Form 1A for construction projects or activities (item 8 of the schedule). In this regard, the regulatory authority may also consult the EAC or SEACs as the case may be.

9.2.9.9 Post environmental clearance monitoring

- For Category A projects, it shall be mandatory for the project proponent to make public the environmental clearance granted for the project along with the environmental conditions and safeguards at the owner's cost by prominently advertising in at least two local newspapers of the district or state where the project is located and, in addition, permanently displaying in the project proponent's website.

- For Category B projects, irrespective of the clearance by MoEFCC/SEIAA, the project proponent shall prominently advertise in newspapers indicating that the project has been accorded environment clearance with the details available on the MoEFCC website.

- The MoEFCC and the SEIAAs, as the case may be, shall also place the environmental clearance in the public domain on government portal.

- The copies of the environmental clearance shall be submitted by the project proponents to the heads of local bodies, panchayats, and municipal bodies in addition to the relevant offices of the government, who in turn has to display the copies for 30 days from the date of receipt.

- It shall be mandatory for the project management to submit half-yearly compliance reports in respect of the stipulated prior environmental clearance terms and conditions in hard and soft copies to the regulatory authority concerned on 1 June and 1 December of each calendar year.

- All such compliance reports submitted by the project management shall be public documents. Copies of the documents shall be given to any person on application to the concerned regulatory authority. The latest such compliance report shall also be displayed on the website of the concerned regulatory authority.

9.2.9.10 Transferability of environmental clearance

A prior environmental clearance granted for a specific project or activity to an applicant may be transferred during its validity to another legal person entitled to undertake the project or activity on application by the transferor, or by the transferee with a written 'no objection' by the transferor, to, and by the regulatory authority concerned, on the same terms and conditions under which the prior environmental clearance was initially granted and for the same validity period. No reference to the EAC or SEAC concerned is necessary in such cases.

9.2.9.11 Operation of EIA Notification 1994, till disposal of pending cases

From the date of final publication of this notification, the EIA Notification number S.O. 60(E) dated 27 January 1994 is hereby superseded, except in suppression[8] of the things done or omitted to be done before such suppression[9] to the extent that in the case of all or some types of applications made for prior environmental clearance and pending on the date of final publication of this notification, the Central Government may relax any one or all provisions of this notification except the list of the projects or activities requiring prior environmental clearance in Schedule I,[10] or continue operation of some or all provisions of the said notification, for a period not exceeding 24 months[11] from the date of issue of this notification.

9.2.9.12 Guidelines for project proponent to prepare EIA/EMP report

- Submit an application to the MoEFCC/SEIAA in Form 1 plus Form 1A (as applicable).
- Refer standard TOR issued by the MoEFCC for the type of industry.
- Collect (monitor) baseline data as per the TOR.
- Based on monitoring, prepare draft EIA/EMP report.
- Submit draft EIA/EMP report to project proponent for applying to the SPCB for conducting public consultation.
- Salient points of public consultation/action to be taken will be incorporated in EIA/EMP report.
- Modified EIA/EMP report to be submitted to the MoEFCC or SEAC as the case may be.
- As advised by the MoEFCC, project presentation to be made for getting environmental clearance.
- Recommendation by EAC/SEAC to the MoEFCC/SEIAA for project approval or otherwise.

9.2.10 National Green Tribunal Act, 2010

The National Green Tribunal Act, 2010 ensures the effective and expeditious disposal of cases relating to environmental protection and conservation

---

[8] Corrected by Corrigendum No. S.O. 1949, dated 13 November 2006
[9] Inserted by S.O. 3067(E), dated 1 December 2009
[10] Renumbered by S.O. 3067(E), dated 1 December 2009
[11] Substituted by S.O. 1737, dated 11 October 2007

of forests and other natural resources, including enforcement of any legal right relating to environment and giving relief and compensation for damages to persons and property and for matters connected therewith or incidental thereto. Any person aggrieved by any award, order of the tribunal may file an appeal to the Supreme Court within 90 days from the communication of the award, order or decision of the Tribunal.

9.2.11 Chemical Accidents (Emergency Planning, Preparedness, and Response) Rules, 1996

The Chemical Accidents (Emergency Planning, Preparedness, and Response) Rules, 1996 provided a statutory back-up for the setting up of a crisis group in districts and states that have major accident hazard (MAH) installations for providing information to the public. So far, 1436 MAH units have been installed in 265 districts. The rules define the MAH installations that include industrial activity, transport, and isolated storage at a site, and handling hazardous chemicals in quantities specified.

As per the rules, the Government of India has constituted a Central Crisis Group (CCG) for managing chemical accidents and has set up an alert system. The chief secretaries of all the states have also constituted standing state crisis groups (SSCGs) to plan and response to chemical accidents in the state. The district collector has to constitute district as well as local central crisis groups (DCG and LCG). So far, 13 states have constituted state, district, and local crisis groups. The CCG is the apex body in the country to deal with and provide expert guidance for planning and handling of major chemical accidents. It continuously monitors the post-accident situations and suggests measures for the prevention of recurrence of such accidents. The MoEFCC has published a state-wise list of experts and concerned officials. The SSCG is the apex body of the state chaired by the chief secretary consisting of Central Government officials, technical experts, and industry representatives and deliberates on planning, preparedness, and mitigation of chemical accidents to reduce the extent of loss of life and property as well as ill health. The SSCG reviews all the district offsite emergency plans for its adequacy. The district collector is the chairman of the DCG, which is the apex body at the district level. The DCG will review all the onsite emergency plans prepared by the occupier of the MAH installations and also conduct a full-scale mock drill of the district offsite emergency plan at the site each year. So far, 1309 units have prepared onsite and 90 district offsite emergency plans.

These rules enable the preparation of onsite and offsite emergency plans, update and conduction of mock drills. Implementation of the PLI Act for providing speedy relief to the victims is also ensured under these rules.

9.2.12 Biomedical Wastes (Management and Handling) Rules, 1998

The Biomedical Waste (Management and Handling) Rules, 1998 regulate the disposal of biomedical wastes, including human anatomical wastes, blood, body fluids, medicines, glasswares, and animal wastes by the health care institutions/hospitals (such as nursing homes, clinics, dispensaries, veterinary institutions, animal houses, pathological laboratories, and blood banks) in cities having population more than 30 lakh or all hospitals with bed strength of more than 500. They were required to install and commission requisite facilities such as incinerators, autoclaves, and microwave systems for the treatment of biomedical wastes by 30 June 2000. All health care facilities were required to provide biomedical waste disposal facilities by 31 December 2002. All persons handling such wastes are required to obtain permission from the appropriate authority. Segregation of biomedical wastes at source has been made mandatory for all institutions and organizations dealing with them. These rules make the generator of biomedical waste liable to segregate, pack, store, transport, treat, and dispose the biomedical waste in an environmentally sound manner. The second amendment to these rules was issued on 2 June 2000.

A steering committee has been constituted to oversee the implementation of the 1998 Rules and the amendment in 2000.

9.2.13 Recycled Plastics Manufacture and Usage Rules, 1999

The Recycled Plastics Manufacture and Usage Rules, 1999 came to be called the Plastic Waste (Management and Handling) Rules, 2011 from 4 February 2011. Under these rules, the use of carry bags and sachets or containers made of recycled plastic or compostable plastic for storing, carrying, dispensing, or packaging food is prohibited. Carry bags and sachets made of plastics can be manufactured only when (i) virgin plastic in its natural shade (colourless), which is without any added pigments, is used and (ii) recycled plastic is used for purposes other than storing and packaging food using pigments and colourant as per IS:9833:1981. Recycling of plastic is to be undertaken strictly in accordance with the Bureau of Indian Standards specifications IS: 14534:1998 entitled 'The Guidelines for Recycling of Plastics'. Carry bags made from compostable plastics shall conform to Indian Standard: IS/ISO 17088:2008. The manufacturer has to print on each packet of carry bags as 'Made of Recycled Material' or of 'Virgin Plastic'. The minimum thickness of carry bags should not be less than 40 microns. And finally, the Plastic Industry Association, through its member units, has to undertake self-regulatory measures. These rules shall be enforced by the concerned SPCB in the states and PCCs in the union territories.

9.2.14 Fly Ash Notification, 1999

The final notification to conserve the topsoil and prevent the dumping and disposal of fly ash discharged from coal- or lignite-based thermal power plants was issued on 14 September 1999 (MoEF 1999b). Under these directives, it is mandatory for every brick manufacturer within a radius of 50 km from coal- or lignite-based thermal power plants to mix at least 25% of ash (fly ash/bottom ash/pond ash) with soil on weight-to-weight basis to manufacture clay bricks or tiles or blocks used in construction activities. Every coal- or lignite-based thermal power plant has to make available ash for at least 10 years from the date of publication of this notification, without any payment or any other consideration, for the purpose of manufacturing ash-based products. Every coal- or lignite-based thermal power plant commissioned subject to environmental condition stipulating the submission of an action plan has to achieve the same within 9 years (15 years for plants not covered by environmental clearance). As per the directive, central and state government agencies, state electricity boards, the National Thermal Power Corporation, and the management of thermal power plants have to facilitate utilization of ash and ash-based products in their respective schedule of specifications. All the local authorities have to specify in their respective building bye-laws and regulations about the use of ash and ash-based products.

9.2.15 Municipal Solid Wastes (Management and Handling) Rules, 2000

The Municipal Solid Wastes (Management and Handling) Rules, 2000 were notified under the Environment (Protection) Act, 1986 on 3 October 2000 (MoEF 2000e). Under these rules, the municipal authority is made responsible for implementation of the provisions of these rules and for any infrastructural development for collection, storage, segregation, transportation, processing, and disposal of municipal solid waste (MSW) and has to comply with these rules given in Schedule I. The municipal authority or the operator of a facility shall apply in Form I for the grant of authorization for setting up waste processing and disposal facility, including landfills, from the SPCB to comply with the implementation programme laid down in Schedule I. An annual report is to be submitted by the municipal authority in Form II to the district magistrate/deputy commissioner, who shall have the power to enforce these rules. MSWs shall be managed as per Schedule II. Disposal of MSWs shall be through landfill as per specifications and standards laid down in Schedules III and IV. The standards regarding groundwater, ambient air, leachate quality, and compost quality shall be followed by the municipal authorities as per Schedule IV. On the basis of Form I, the SPCBs/PCCs shall issue authorization in stipulating compliance criteria and standards as specified in Schedules II, III, and IV. Authorization shall be valid for a given period. The CPCB and SPCBs/PCCs will review the implementation of standards and guidelines. The SPCBs/PCCs shall submit regular reports to the CPCB in Form IV regarding the implementation of the rules by 15 September every year. The CPCB shall prepare the consolidated annual review report on the management of MSWs and submit to the Central Government along with its recommendation before 15 December every year. Municipal authorities have to submit an accident report in Form VI in the case of an accident during collection, segregation, storage, processing, treatment, and disposal facility or landfill site or during the transportation of such wastes.

9.2.16   Batteries (Management and Handling) Rules, 2001

The MoEFCC issued the Batteries (Management and Handling) Rules, 2001 to control the hazards associated with the backyard smelting and unauthorized reprocessing of lead–acid batteries. The lead–acid batteries are widely used in inverters and automobiles such as cars, trucks, buses, three-wheelers, two-wheelers.

As per the rules, it is mandatory for the battery manufacturers, importers, assemblers, and re-conditioners to collect old batteries, on a one-to-one basis, against the sale of new batteries. The batteries so collected have to be sent to recyclers, registered with the MoEFCC for recycling and processing, using environmentally sound technology, unless battery manufacturers themselves have such recycling facilities. Registration is accorded by the MoEFCC to only those units that have installed environmentally sound technology for processing lead–acid batteries, pollution prevention systems, and suitable arrangements for waste disposal. As a result, backyard smelting of lead–acid batteries with attendant lead emission to the atmosphere, discharge of acid into open ground sewers, and loss of load due to poor recovery (30%–40%) will come down substantially.

Manufacturers are required to set up collection centres for the collection of used batteries. Collection centres can be set up either individually or jointly. As such, small-scale manufacturers can set up collection centres jointly or make use of collection centres set up by others. Batteries sold to bulk consumers such as central/state government departments, state road transport undertakings, and original equipment manufacturers such as automobile manufacturers have been excluded from the obligation for collection. Bulk consumers/auctioneers can auction used batteries only to the registered recyclers or processors registered with the MoEFCC and small-scale manufacturers are at liberty to procure recycle lead from registered recyclers. As a result, middlemen and backyard smelters would be debarred from participation in any auction within the country. Also manufacturers have to incorporate a suitable provision for buyback in the case of bulk sale of batteries by the manufacturers to bulk consumers.

Dealers have also been assigned responsibility for collection. The auction of used lead–acid batteries by bulk consumers or auctioneers can be made only in favour of registered recyclers. As a result, the supply of lead for backyard smelting will be reduced substantially. Importers as well as domestic manufacturers have been brought under the provision of these rules and are placed on the same footing as far as responsibility for collection is concerned.

Manufacturers, assemblers, re-conditioners, importers, recyclers, auctioneers, users, and consumers are required to submit half-yearly returns to the SPCB, which has been designated as the prescribed authority. The forms have been designed to enable easy verification of responsibilities fixed for everyone under the rules. A collection schedule has been prescribed providing for gradual enhancement of the percentage of the batteries to be collected and achieve a level of 90% collection from the third year as mentioned in Schedule I. Batteries have been categorized to ensure that batteries collected are similar to that of the batteries sold. Since the MoEFCC already has a scheme for registering pre-processors of used lead–acid batteries, only those re-processors who have not applied to the MoEFCC already will be required to apply for registration.

9.2.17   Prevention of Cruelty to Animals Act, 1960

The Prevention of Cruelty to Animals Act was passed in 1960 to prevent unnecessary pain or suffering to animals and to amend the laws related to the prevention of cruelty to animals. After the enactment of this act, the Animal Board of India was formed for the promotion of animal welfare.

9.2.18   Wildlife (Protection) Act, 1972

The Government of India enacted the Wildlife (Protection) Act, 1972 with the objective of effectively protecting the wildlife of this country and to control poaching, smuggling, and illegal trade in wildlife and its derivatives. The act was amended in January 2003, and punishment and penalty for offences under the act have been made more stringent. The ministry has proposed further amendments to the law by introducing more rigid measures to strengthen the act. The objective is to provide protection to the listed endangered flora and fauna and ecologically important protected areas.

9.2.19   Biological Diversity Act, 2002

The Biological Diversity Act, 2002 is a federal legislation enacted by the Parliament of India for the preservation of biological diversity in India. It provides a mechanism for the equitable sharing of benefits arising out of use of traditional biological resources and knowledge. The act was passed to meet the obligations under the Convention on Biological Diversity, to which India is a party.

9.2.19.1   *Biodiversity and biological resource*

Biodiversity has been defined under Section 2(b) of the act as "the variability among living organisms from all sources and the ecological complexes of which they are part, and includes diversity within species or between species and of eco-systems". The act also defines biological resources as "plants, animals and micro-organisms or parts thereof, their genetic material and by-products (excluding value-added products) with actual or potential use or value", but does not include human genetic material.

9.2.19.2   *National Biodiversity Authority and state biodiversity boards*

The National Biodiversity Authority (NBA) is a statutory autonomous body, headquartered in Chennai, under the MoEFCC established in 2003 to implement the provisions under the act. State biodiversity boards (SBBs) have been created in 28 states along with 31,574 biological management committees (for each local body) across India.

Functions
- Regulation of actions prohibited under the act
- Advise the government on the conservation of biodiversity
- Advise the government on the selection of biological heritage sites

- Take appropriate steps to oppose grant of intellectual property rights in foreign countries, arising from the use of biological resources or associated traditional knowledge

Regulations

A foreigner or non-resident Indian, as defined in Clause (30) of Section 2 of the Income Tax Act, 1961, or a foreign company or corporate body needs to take permission from the NBA before obtaining any biological resources or associated knowledge from India for research, survey, and commercial use. Indian citizens or corporates need to take permission from the concerned SBBs.

The results of research using biological resources from India cannot be transferred to a non-citizen or a foreign company without the permission of the NBA. However, no such permission is needed for publication of the research in a journal or seminar, or in the case of a collaborative research made by institutions approved by the Central Government.

No person should apply for patent or other forms of intellectual property protection based on the research arising out of biological resources without the permission of the NBA. While granting such permission, the NBA may make an order for benefit sharing or royalty based on the utilization of such protection.

Benefit sharing

Benefits arising out of the usage of biological resources can be shared in the following manner:
- Joint ownership of intellectual property rights
- Transfer of technology
- Location of production, research development units in the area of source
- Payment of monetary and non-monetary compensation
- Setting up of venture capital fund for aiding the cause of benefit claimers

Penalties

A person violating the regulatory provisions may be "punishable with imprisonment for a term which may extend to five years, or with fine which may extend to ten lakh rupees and where the damage caused exceeds ten lakh rupees such fine may commensurate with the damage caused, or with both". Any offence under this act is non-bailable and cognizable.

9.2.20   E-waste (Management and Handling) Rules, 2011

The E-waste (Management and Handling) Rules, 2011 were notified by the Government of India, vide number S.O. No. 1035 (E), dated 12 May 2011 in the Gazette of India under the powers conferred by sections 6, 8, and 25 of the Environment (Protection) Act, 1986 and came into effect from 1 May 2012.

These rules apply to every producer, consumer, or bulk consumer involved in the manufacture, sale, purchase, and processing of electrical and electronic equipment or components detailed in Schedule I of the rules.

All bulk users of electrical and electronic equipment, such as central or state government departments, public-sector undertakings, banks, educational institutions, multinational organizations, international agencies, and private companies registered under the Factories Act, 1948 and Companies Act, 1956, come under the ambit of these rules. Moreover, these rules also define the role and responsibility of all bulk users, collection centres, consumers, dismantler, recycler, producer, and transporter, who may be involved in handling, generation, collection, reception, storage, segregation, refurbishment, dismantling, recycling, treatment, or/and disposal of e-waste.

Thus, to ensure compliance to the rules, authorities have to ensure:

- Collection, segregation, and channelization of e-waste generated.
- Authorization from the pollution control board (PCB) in accordance with the procedures under Rule 9 of the rules.
- Maintaining records and inventory of e-waste generated/handled/ disposed in Form 2 and make them available to the PCB during inspections as and when required.
- Filing annual returns in Form 3 to the PCB on or before 30 June following the financial year to which the return relates.

10
Education for Environmental Awareness

10.1 INTRODUCTION

Students of environmental education should recognize that conscious and formal efforts are necessary to ensure education and awareness for all sectors of the society. This chapter discusses the history of and justifications for environmental education. It also includes simple suggestions and activities that people can incorporate in their daily life, at home or in a more academic setting, and their community to foster the ongoing environmental awareness.

10.2 HISTORY AND PURPOSE OF ENVIRONMENTAL EDUCATION

When one considers the environmental issues affecting local and global development and the need for public participation and understanding, one realizes the importance of helping people of all ages nurture respect for the environment. Therefore, environmental education has fast become a standard and a 'required part of all' formal educational programmes around the world.

According to Shet (2003) the curriculum for environmental education has been implemented rather sluggishly in India. According to her, "Thanks to a two-year study that identified the gaps and anomalies in environmental education in India, 800 schools now have a new and improved syllabus that promotes an understanding of environmental issues. More than 100 schools in the state of Maharashtra, and 700 more around India, now have a syllabus that aims to improve children's understanding and knowledge of the environment."

An understanding of human interconnectedness and dependency on nature is instilled at an early age in people growing up in farming communities or villages. In India, as discussed in previous chapters, this still occurs for about 70% of the population. From the dawn of human ancestry to the Industrial Revolution, this was the reality for all humanity. In such times, there was no need for a formal study of 'environmental education'. Life and existence depended on an intimate knowledge of nature. While human activity could upset the balance of nature at times, human technology and activity never threatened as much as it does today.

It is no wonder that the modern environmental education and conservation movements in the USA started during the Industrial Revolution, when and where this deep connection with nature began to get lost. More and more people moved to the city and purchased products that they needed to live, items produced and transported by burning coal. The Audubon Society, an environmental organization formed in 1896, began a protest movement, chiefly organized by women who were appalled at the use of stuffed birds on women's hats. The use of whole birds on hats was driving these exotic birds to extinction. So, the women took action and were able to ban the use of exotic birds. The organization started and still works with a focus on bird conservation.

John Muir, founder of the Sierra Club, also foretold the loss of habitat, humanity, and enjoyment of nature in 1892 as he feared the outcome of the Industrial Revolution. President Theodore Roosevelt, known as the Conservation President, preserved 230,000,000 acres of national forests during his presidency (1902–09).

Thus, in the early 1900s, many began to see that an appreciation and understanding of nature should be taught and encouraged in an urban and modern world. Since these early awakenings, this awareness has only continued to grow. During the 1950s, the use of the term 'environmental education' began, and with the publishing of Rachel Carson's *Silent Spring* in 1962, the modern environmental movement was considered born. The first view of the earth from space came in 1969; the first Earth Day was observed in 1970; the US Clean Water Act was passed in 1972; the Endangered Species Act was passed in 1973; and the Environmental Education Act was passed in 1990, which mandated that the US Environmental Protection Agency provide national leadership to increase environmental literacy. The five components that make up the purpose of environmental education are as follows:

1. Awareness and sensitivity to the environment and environmental challenges
2. Knowledge and understanding of the environment and environmental challenges
3. Attitudes of concern for the environment and motivation to improve or maintain environmental quality
4. Skills to identify and help resolve environmental challenges
5. Participation in activities that lead to the resolution of environmental challenges

As urbanization and disconnect from nature become greater around the world and technology increasingly shapes our world and mindset, the need for formal environmental education is clear. The United Nations Conference on the Human Environment in 1972, as well as the United Nations Educational, Scientific, and Cultural Organization and the United Nations Environmental Programme in 1977 created the Tbilisi Declaration, provided rationale and guiding principles for environmental education. The more recent United Nations Conference on Environment and Development in Rio de Janeiro (Brazil) in 1992 called for education oriented towards sustainable development. As Richard Louv recently noted in *Last Child In The Woods*, the emotional and psychological health of children may even be in danger without experiential activities in nature.

Challenges and controversies are abound in teaching environmental education. Some feel that environmental education can become propaganda for certain political agendas and not a way to stimulate independent thinking based on science. Some find that there is lack of time to teach the topic due to other demands. Whether it is taught or used in the classroom depends on the interest of administrators or individual teachers in many locations where there is no state mandate for environmental education. India has some standards connected to environmental education, but it does not require a specific K-12 environmental education curriculum. If teachers are raised in the city, they themselves may not be comfortable with nature and feel more comfortable staying in the classroom. Many teachers in other subject areas feel environmental education should only be the domain of science teachers. General apathy towards the environment and a feeling that small actions will make no difference in stopping large global issues also discourage some from studying the environment.

Finally, in urban areas where environmental education is so desperately needed, it is often difficult to find access to wildlife and nature to strengthen an environmental education curriculum. There may be logistical difficulties in monitoring large groups of children outside and no proximity of natural areas to schools, as well as liability for injury and insect bites. Sometimes the fear and dislike of urban children for nature due to non-exposure hinder teachers more. The costs and safety concerns of trips to distant locations are obstacles too, especially in impoverished neighbourhoods and slums.

While these are real challenges, consider the following justifications for basic experiential and environmental education:

- Psychologically, emotionally, creatively, and spiritually there are reasons to be in nature to promote connectedness, tranquillity, and a sense of awe, wonder, and curiosity about life.

- Most environmental educators and regular educators agree that all environmental problems and issues are better understood if children have had first-hand exposure to and connection with wildlife and natural systems (carbon, water, and geological cycles). According to a study by Louise Chawla of the Kentucky State University, "most environmentalists attributed their commitment to nature to many hours spent outdoors in a wild or semi-wild area and an adult who taught respect for nature". Simply being told about the problems and made to fear and be overwhelmed by them, especially distant ones, did not seem to have as much influence and was echoed by studies on children. David Sobel writes, "If we want children to flourish, to become truly empowered, then let us allow them to love the earth, before we ask them to save it."

- Once people have had the chance to bond with, connect to, and value nature, they can be guided to scientifically study it and figure out where problems exist and suggest possible solutions.

- Using an environmental focus can be an excellent way to teach the process of science and at the same time maths, reading, writing, social studies, and even art.

- Once students have been turned on to learning through environmental education experiences/projects, they feel motivated, empowered and will hopefully retain these ethics and possibly pursue careers in this field or at least live a life that considers these issues.

- Children who live in extreme poverty most likely have never travelled to exotic wild places or been allowed to just play in nature. However, in both developing nations and impoverished inner cities, environmental issues (lack of water, air pollution, toxic waste dumps) almost always disproportionately impact the poor.

It is, thus, clear that all people, especially children from urban areas and disadvantaged backgrounds, need experiences in nature and should engage in activities that will help them connect on a more spiritual manner. This intrinsic love for nature, combined with formal academic studies of the environment, can hopefully foster people to act and engage in projects, careers, and politics that foster sustainability for people and the planet.

This education need not happen only in school. The urgency of addressing environmental issues cannot wait for mandated educational curriculums to catch up. Environmental education can and should be incorporated in many settings. Families, youth, women, religious and civic groups can easily conduct these activities. The next section provides some basic ideas and suggestions for getting involved in developing awareness, sensitivity, knowledge, attitudes, and skills and then perform actions that will help address various environmental challenges. While these simple steps are effective, especially if everyone in a society is practising them, more importantly they help develop knowledge, sensitivity, and ethics for constant environmental concern and preservation. These small steps are, however, not the only goal. Hopefully, success with developing these habits will lead to active citizen participation in broader community decision-making, which will have even greater impact on the environment. Figure 10.1 shows how one can achieve sustainability at home by judicious use of the resources.

Figure 10.1 A local New Delhi woman uses her home to teach others about sustainability. She composts, uses a solar cooker, and dries cow dung as fuel for cooking on a gas stove.
Picture courtesy Lynn Tiede

10.3 ACTIONS AND IDEAS FOR ENVIRONMENTAL SUSTAINABILITY AT HOME

Where and how one lives affect the environment, and analysis helps understand the impact. The energy used to heat, cool, run appliances, and facilitate transportation impacts pollution levels and climate change. Water usage and sewage treatment as well as the chemicals used in cleaning and washing impact aquatic ecosystems and freshwater availability. All consumer and household products involve activities such as mining, industrial manufacturing, and transportation and can emit toxins into the air and ground during usage or disposal. Indeed, adopting a mentality of water and electricity conservation, reuse, and preferring naturally made products can lessen this impact.

10.3.1 Analysis of the Environmental Impact of Home

- What is the size of the household in which you live, whether house or apartment, both the building and land (if any)? How many rooms are there? Is the property owned or rented?
- Make a sketch of the floor plan of the property, the layout of the rooms, and any attached land.
- How many people live there and what ages are they?
- What transportation does the household use, and how much on a regular basis versus the occasional? Consider various modes— walking, bicycle, public transportation by bus, subway, or train, taxi, automobile, airplane.
- What resources are used?
- What is the electricity supply? How many circuits are there and with what rating? How much electricity is used in a month? How reliable is the electrical service? If not electric, what is the source of lighting?
- Is the household heated or cooled during any part of the year? How much incremental energy is used for that? What system or source of energy is used?
- What is the source of water and how much is used in a month? What is the quality and reliability of the supply?
- How is food cooked and with what fuel or energy source? What is the source of the food? Where was it purchased, and how and where was it grown? Is any food home grown?
- What consumer, cleaning, and household products are used? What types of chemicals are these products made of? Is care taken to

find the least toxic options and to reuse and buy only what is essential?

- How is waste disposed of—both sewage and trash? Is any waste recycled or donated?
- For all utilities, what is the cost of each in an average month—water and electricity, transportation, fuel, and waste disposal?
- Are there any green features, such as solar panels or heaters or rainwater harvesting?
- How is free and leisure time used? Does the family spend time in nature or conduct activities that are not reliant on energy and electronics usage?
- After gathering your data, can you suggest ways to reduce the impact of your household on the environment to make it greener?

Following are some other actions you can take after your investigation:

- **Reducing energy use**
  - Turn off lights and equipment that are not in use, especially computers. Unplug other appliances when not in use. Put all items on a power strip and turn this off when not in use. It is estimated that 'phantom energy' (energy consumed when items are plugged in) could save 5%–10% of one's energy use.
  - Install compact fluorescent lamps where possible.
  - When making new purchases, make sure products are energy efficient or have sleep functions.
  - Carpool, ride a bike, or walk if possible.
  - Reduce the amount of time you watch television, listen to music, or use your computer. Go outside instead. Observe your surroundings.
- **Water conservation**
  - Alert individuals to leaking faucets. Low-flow shower heads and faucets are available.
  - Take small tub baths or very short showers.
  - Investigate whether your home can install a rainwater harvesting system.
  - Do not leave water running while brushing your teeth.
- **Air quality**
  - Make sure you have plants inside. According to a National Aeronautics and Space Administration (NASA) study, good plants for filtering the air include reed palm, dwarf date palm,

Boston fern, Janet Craig Dracaena, English ivy, Australian sword fern, peace lily, rubber plant, and weeping fig.

- Avoid chemical cleaners, paints, and plastics in the home to avoid gassing.

• Material goods consumption and waste management
- Purchase environmentally-friendly and fair-trade products made with recycled or natural content.[1]
- Reuse the empty sides of discarded paper for printing.
- Recycle or reuse all materials as possible, including bottles, magazines, envelopes, and newspapers. Aim for zero waste. Take care when disposing of cleaning water, fluorescent light bulbs, batteries, or other toxic materials. Electronics, toner cartridges from printers, and even sneakers can be recycled in some countries. Is this happening in India?
- Create a home compost bin. Compost any spoiled food items.
- Think twice before buying or asking for games, clothes, and shoes that you might not need.
- Bring a reusable bag when you go shopping.
- Purchase natural cleaning products or help your parents make environmentally friendly cleaning products. Natural oils, vinegar, baking soda, and pure soaps are less toxic for the home. Avocado, egg whites, yogurt, mango, pineapple, and sugar can all be ingredients in cosmetics and beauty products.
- Shred important documents and use the paper for packaging materials.
- Recycle toner cartridges from computers and copiers.
- Use small post-it fax notes rather than a large fax cover sheet.
- Use rechargeable batteries for battery-operated equipment.
- Buy in bulk and shop for products with less packaging.

• **Food**
- Purchase organic and locally grown foods as much as possible. Maintain a home garden if you can. A small area or containers can even produce a small amount of vegetables.
- Reduce the amount of bottled drinks, packaged food, and junk food you buy. It is healthier for you and creates less trash. Opt for traditional street food from local vendors.

---

[1] Websites of some companies that offer environmentally-friendly products in India are www.taraprojects.com, www.biotique.com, ww.fabindia.com, www.fairtradeforum.org, and www.sadhna.org

- Maintain vegetarian customs; meat production is environmentally taxing since more land, water, and energy are used. Rainforests are also often destroyed for cattle grazing.
- **Leisure and free time**
  - Donate toys, clothes, books, and other things you no longer need.
  - Ask for and give gifts that are earth friendly—nature guides, nature study computer programs, binoculars and microscopes, rechargeable batteries/battery charger, school supplies and stationary made from recycled paper, books on nature and environmental issues, trip to a special natural place.
  - Do not always chose activities using technology to entertain yourself. Take a walk outside, go to a park, read about nature, and plant a tree or flower plant. Conduct a home audit to find ways your whole family can reduce usage of water, energy, and toxic non-disposable material goods.

10.4 FOSTERING ENVIRONMENTAL EDUCATION IN AN ACADEMIC SETTING

- Consider selling responsibly made products as special fundraisers. Recycled paper notebooks, pencils, binders, and pens are available. T-shirts are available from recycled or organic cotton.
- Host special events that emphasize responsibility. Sponsor collection of old clothes, sneakers, or old greeting cards if these items can be recycled or reused in your community. Host parties, but require participants to bring an item that can be reused or recycled.
- Make double-sided copies whenever possible.
- Consider careers that incorporate nature and environmental protection. Scientists, lawyers, political leaders, and educators can be directly involved with these issues, but most careers can incorporate green ethics. Study leaders in green businesses, green architecture, organic farming and get more ideas of how your future can incorporate a respect for the environment.

10.5 ENVIRONMENTAL EDUCATION IN YOUR COMMUNITY

As your skills and understanding of how your daily actions can impact the home and academic settings grow, the next step is to broaden personal and community awareness for your entire locality and nation. General ideas for cultivating this awareness are as follows:

- Increase your knowledge of nature and environmental issues in your community. Many institutions offer classes on environmental topics. Public libraries have lots of resources to stay informed.
- Become a volunteer. You can help monitor birds and other wildlife, plant trees, clean parks and streets of litter, and lead tours and talks for younger kids.
- Join a local environmental organization. Learn about local environmental issues and see what efforts you can help with.
- Choose an environmental issue in the country as a whole or in your state or community, and apply a reasonable method of activism about that issue, such as a letter-writing campaign. Send petitions and letters to leaders.
- Read newspapers. Be aware of changes taking place in your community, including development projects. Consider and think about effects of these projects on the environment. Posters and billboards can be helpful in creating public awareness (Figure 10.2).
- Do the right thing when you are outside in nature:
  - **Stay on the paths:** Walking on marked paths keeps soil healthy and reduces erosion.
  - **Leave nature where it is:** Removing plants, feathers, and birds' nests is illegal for many species and is destructive to the environment.
  - **Do not litter:** Place garbage in trash cans or recycle or reuse it if possible. Animals who eat litter or use it in other ways can die.

The final step would be to conduct a community analysis. As this involves more details, several people working together would be ideal. The assessment should be divided into two parts—assessing the current situation and problems, followed by analysis and action.

1. **Assessing the general situation:** What is the general layout—buildings and open areas, streams and rivers, housing and businesses and factories, streets and roads, railroad, and airport? Is there a public transport system? What is the source of water and electricity for the community? What waste sources are there, point and non-point, and how is waste disposed of? Is there provision for recycling or other 'green' efforts? How can various activities be categorized as green versus non-green (or the degree of greenness)? For example, compare the amount of energy needed to travel per kilometre on foot, bicycle, driving in different vehicles,

Figure 10.2 Posters and billboards can be effective methods for teaching the public how to green their community

Picture courtesy Lynn Tiede

public transport of various kinds, and by air. Calculate the 'carbon footprint' of various activities, from transport to heating and cooling to food production and transportation. This means how much carbon (as carbon dioxide) is generated by various activities. Similarly, consider the 'water footprint' or overall water use by the community.

2. **Surveying environmental problems:** What pollution is there: air, water, and land? Are there environmental diseases such as malaria? Are the water and electrical supplies reliable? Is waste disposal adequate and dependable? Is flooding a concern?

3. **Analysis:** What can be done to make the community more 'green'? Which problems are the easiest to solve? Which cost the most and least in money and effort?

10.6 CONCLUSION

Environmental education must become part of all levels of formal education, primary through college, as well as public awareness campaigns in the modern world. While governments, school districts, and academic institutions explore and debate this, individuals can take steps together and on their own to become more aware of their environment and to preserve it.

11

Projects and Activities

11.1 INTRODUCTION

Academic and field projects can encompass an entire range of issues. The aim of environmental education is to help people become environmentally aware, knowledgeable, skilled, and dedicated citizens, who can work individually and collectively to defend, improve, and sustain the quality of the environment on behalf of the present and future generations of all living beings. For this, people need to build perceptual awareness, knowledge, environmental ethics, and action skills through experiences and hands-on investigation.

A person's age and prior learning, as well as the needs of the community and the availability of resources, should guide the issues approached and the type of activities that are suitable. However, a comprehensive investigation would include all components of the environment (air, water, plants, animals, energy, land use/green space, consumer responsibility, and waste management) and all the goals of environmental education.

11.2 NATURE OBSERVATION AND WILDLIFE APPRECIATION ACTIVITIES

11.2.1 Ecology Observations and Investigations

If you live in an urban environment, there is a good chance that you are a bit nature deprived and lack first-hand experiences and observation of nature. So the first step would be to conduct some research to create a list of what natural areas are available nearby. One needs to study local habitats so that one can really investigate nature in a hands-on manner and if possible see the results of an action. If one simply studies about rainforests or animal extinction on another continent, it is harder to feel truly connected even if one adopts an acre of the rainforest. There is still disconnect. Do a survey to see what types of resources may be available nearby. Here are some ideas:

- Local community gardens/farming fields
- Protected natural areas and nature preserves
- Local city parks and waterfronts
- Non-governmental organizations (NGOs)/universities or other organizations may offer special programmes that create people's interest in nature or they may possibly make available for use, such as microscopes, nets, collection jars.

Once you identify the options, go to these local natural areas. Keep a journal, draw and observe nature, and identify wildlife. Visit during different seasons so that you can see changes. Formulate questions and create your own scientific inquiries. You can study such topics as plant structures, insects, erosion, soil composition, rainfall averages, and identification of species.

11.2.2 Studying Birds in the Wild

Studying birds is an excellent way to achieve many of the goals of environmental education. Birds play a special role in every ecosystem as they are both prey and predators of other species. Birds keep the populations of insects and small mammals in check. They also help pollinate flowers and transport seeds. They are present everywhere, and they are a stepping stone to engaging people in the rest of the natural world and environmental issues. They are an easy way to teach interdependence and the concept of adaptations for different habitats. Their variety of shapes, colours, sounds, and sizes are fascinating. Great poetry, art, literature, myths, and symbolism have been created around birds in many cultures.

Birds are sometimes called the "avian litmus paper"; the health of their population mirrors the soundness of the broader ecosystems and environment in which they live. Scientists like to watch for trends in the decline of bird populations and can get leads on problems that may ultimately affect humans. Birds are smaller and thus often show reaction to chemicals faster than humans. The use of DDT was banned for this reason. However, DDT is still used in some countries, and so birds travelling to those parts of world still die. In the United States, the usage of other pesticides in farms and lawns is killing them. Many other environmental issues can be highlighted when bird populations are threatened.

Finally, the fact that birds migrate makes them very interesting to study for teaching the concept of think globally, act locally. Destruction of bird habitats in the rainforests of South and Central America as well as the United States is causing many species to die (it is estimated that

250 species have died because of South American rainforest destruction). Any damage to their habitats affects their survival rates—from pollutants; killing plants and animals they need for food; and development projects that reduce their habitat, such as mining, oil extraction, cell phone towers, and building of skyscrapers. Studying birds is a fundamental way to assess the impact of human actions on the environment.

Data collection projects allow people to actually go out and count birds. Data collected by citizens can be sent to central databases and used by scientists to find trends. This is called "citizen science". The following programmes relate specifically to birds and are good examples. Some allow international participation and can serve as templates for projects to start where you live.

- **Great Backyard Bird Count:** These events are organized by the Audubon Society. Each group is responsible for a certain natural area. The group counts the birds they see in that environment on a particular day and sends the data. This project is 100 years old.[1]

- **FeederWatch:** Participants can set up feeder that they observe. They create experiments and scientific inquiry to learn about the birds they see. Information is submitted yearly and the best is published. This is a great way to teach the process of scientific inquiry.[2]

- **Celebrate Urban Birds:** This activity engages participants (especially children) in learning about birds where they live through all types of studies.[3]

- **Birds in Forested Landscapes:** This Cornell University project involves citizen scientists in monitoring specific species in forested areas.[4]

11.3 CREATING AND PRESERVING HABITATS FOR WILDLIFE AND GREEN SPACE

Preservation and creation of green space are related to wildlife studies. As development and population grows, these areas must be preserved or often rebuilt in urban areas. There are many reasons to focus on the overall amount and health of green space in one's community. One's psychological, spiritual, and emotional health needs are important

---

[1] Details available at http://www.birdsource.org

[2] Details available at www.birds.cornell.edu/pfw

[3] Details available at http://celebrateurbanbirds.org

[4] Details available at www.birds.cornell.edu/bfl

reasons for preserving green spaces. These are the places to go for hiking, observing nature, and finding solace. Parks are habitats for living animals and plants. If a society wants these species to survive, green spaces where they exist must be maintained. For example, during migration, birds fly for very long distances. They must stop on their journeys to re-fuel. Because of urbanization, natural habitats have been destroyed. Therefore, small plots that provide food are very important to birds and other animal species. When considering endangered animals such as tigers, the conservation of larger preservations is critical to the survival of this magnificent species.

Also, keep in mind that parks and nature reserves are made up of primarily trees and green plants, which provide mechanisms to remove carbon, and filter other toxins, and produce oxygen, which all humans need to survive. Trees and plants absorb water and prevent run-off. They also cool places, which reduces energy use for cooling in hot climates.

Threats to green spaces are owing to the following reasons:

- Economic development purposes
- Recreational uses (sports, hiking, camping), which create compaction, litter, erosion, and flooding
- Lack of funds for the maintenance of green spaces

Get involved in the following activities as you seek to preserve habitats for plants and animals and understand their importance in your locale:

(i) Creating a garden near a home, public place, or school or even on a rooftop can provide one with much learning and appreciation of nature. Seeking to grow native plants and make a garden into a bird and wildlife habitat makes the project even more important. This task entails researching bird-friendly plants, building and making birdfeeders, learning about soil, water, climate conditions that promote growth, and actually growing plants. If one is successful in attracting animals and insects, observation of animals will be easier.

(ii) Work with local green spaces to plant, clean up, or inventory wildlife. Perhaps make it more bird or butterfly friendly. Do a clean-up of some area such as a park or other public open space or of a street or walkway or of a stream. Summarize what you have found by weight or volume or type of material or combination of those. Take pictures and publicize the result or invite the press.

(iii) Inventory, measure, and map all green spaces in your community. Calculate green space available per person and trees per person.

(iv) Go with a group on field trips to a site of environmental interest. The choices depend on what is available in your area, but it could be natural areas and nature preserves, developments, mines, waste disposal sites or landfills, recycling centres, waste treatment facilities, power generation stations, chemical plants, or even "dumps" where disposal was done illegally or improperly. Before such a trip, have some idea of what to expect and follow all guidelines. Record your observations and ask questions of guides. Formulate ways to establish relationships with these entities and work for change to preserve green space or minimize impact on green spaces where needed.

(v) Investigate the web for other resources for green space and habitat preservation and creation.

11.4 WATER QUALITY AND CONSERVATION

All major cities and nations in the world are facing problems regarding the depletion of groundwater, water quality, and pollution. The following are activities for research and investigation:

- **Watershed preservation and protection:** Where does your water come from? How is the watershed or the source of drinking water being protected from pollution and development? What are the costs of building filtration systems and how will this affect the cost of water?

- **Water quality:** Environmental agencies must perform tests to ensure that water is free from microbes, inorganic contaminants, pesticides, herbicides, organic chemical contaminants, and radioactive contaminants. How does this work in your locality? How is bottled water produced and tested?

- **Lead contamination:** Water can pick up lead from solder, fixtures, and pipes found in the plumbing of some buildings and homes. People who live in many older buildings may have a greater risk. See if you can order a lead testing kit and do an analysis.

- **Pollution of waterways, watersheds, and ground aquifers:**
  - *Point sources (easily identifiable sources):* Chemicals and heavy metals from factory wastes; fertilizers, pesticides, and herbicides from farming; animal waste and by-products from dairy, poultry, hog, cattle farms, and production facilities; and a toxic brew of chemicals from city sewers and landfills, as well as leakages from septic tanks and gas storage tanks, are the main sources of pollution. Investigate major threats in your

community and efforts to educate and reduce these threats. Start a campaign yourself.

- *Non-point sources—pollution from dispersed and hard-to-identify locations that enter the waterways:* During significant rainfall, sewers overflow, carrying combined household and industrial wastes directly to waterways. Materials discarded on streets or sidewalks can be carried directly to waterways. Paved areas or areas cleared of vegetation create quicker run-off. Individual use of pesticides and fertilizers and improper disposal of chemicals and cleaning materials and expired medications all contribute to pollution in our waterways. Investigate major threats in your community and efforts to educate and reduce these threats. Start a campaign yourself.

• **Controversy on fluoride treatment:** Some believe it is not necessary to add fluoride to water. Fluoridation may be worthless, creates discolouration of teeth (dental fluorosis), and is harmful when swallowed. Calcium and good nutrition are more important for preventing tooth decay. Investigate the status of fluoridation in India and issues surrounding this controversy.

• **Conservation of water and drought:** Only 2% of the world's water is drinkable. In many places, it comes from underground aquifers that replenish slowly. In certain geographical areas, reservoirs do replenish naturally, but it is disrupted when there is a shortfall of rain or snow. With global weather patterns changing, droughts are becoming huge threats. For limiting our use of water, shorter showers and turning off faucets, as well as reducing consumption, changing manufacturing and farming processes, and changing to a vegetarian diet, are all water conservation practices. Research these types of practices and see if there are any education efforts in your community to improve water conservation. Start some, if not all.

- You can also calculate how much water you and your family are using at home. Search for drips/leaks, and strategize methods to conserve better.[5]

- Consider rainwater and how it is handled. Many parts of India have too much water during the monsoon season and sometimes there is not enough water in other periods of the year. Water harvesting is a general term, which is one way of addressing the problem of uneven distribution of water during

---

[5] Details available at www.waterfootprint.org/?page=files/home

the year. The collection could be from roofs into barrels. The water could be directed into rain gardens or a pond. Water will also infiltrate into the ground and maintain the water table so that wells do not go dry. Expand on this in ways appropriate to your house or community and part of the country. What water supply problems are a concern in your community and what water harvesting methods are best suited to the climate of your town and area.

- **Fish / seafood and human consumption:** Toxins in waterways concentrate in seafood flesh and can harm humans when ingested. In addition, overfishing and unsafe fishing practices of many species are destroying the biodiversity, habitat, and balance of our oceans and waterways. Research the status of these issues in India.

- **Sewage treatment plants:** Sewage treatment plants are necessary and must follow laws regarding their operation, but humans must be concerned about the gases, sludge, and nitrogen-rich wastewater released by them. Investigate how your municipality's wastewater is treated and any concerns in your community.

- **Revitalization / preservation of waterways:** Conservation and health of our waterways are crucial to the survival of species and recreational enjoyment. Research the efforts to address the problem of preservation of rivers, lakes, wetlands, or other waterways in your area. Research ways by which these places may be under threat from development or human overuse. Studying what is happening with the River Ganga would be a great start.

- **Access to waterfront:** Access to waterfront for recreation is limited in many areas of the world. Development must include this option, but it is often overlooked. Investigate whether your locality or other cities in India are addressing this issue through plans for the revitalization of waterfronts.

- **Water rights:** Bottled water is big business. Many national corporations are on a quest to purchase rights to aquifers and sell bottled water worldwide, sparking much conflict, depletion of water for local municipalities, and water shortages. Some fear many future wars will be over water as worldwide supply diminishes. Research how this issue is affecting India.

- **Climate change:** Scientific evidence indicates that world temperatures are increasing. Many of the effects can be projected, but the extent and timing remain somewhat uncertain. This climate change is causing sea levels to rise. Coastal areas across the globe could be severely affected if this trend continues. See what places in India are under threat.

- One effect that has been noticed is melting of glaciers. If the glaciers disappear in the Himalayas, the seasonal flow in many of the major rivers of India will vary even more, from very little flow to more extreme flooding. Research to what extent this has already happened. If you live near a river that originates in the Himalayas, ask older people if they have noticed any difference in the flows now compared to decades ago.

- If the oceans were to rise as much as 1 m, a projection possible if there is substantial melting of the ice caps on Greenland and Antarctica. What areas of India and Bangladesh would be most affected or even inundated. Remember that areas do not have to be normally under water to be affected, as storm surges from a typhoon carry well inland. Also, current coastal areas such as mangroves that can buffer wave action may not survive in deeper water.

- What are some of the scenarios both for climate change and for the possible effects of the change? See Clean Air Cool Planet (www.cleanair-coolplanet.org)

- **Water and wildlife:** In the United States, the National Audubon Society, National Resources Defense Council, Greenpeace, and Sierra Club monitor and work to protect local waterways, wetlands, rivers and oceans, and the wildlife that inhabit them. Since 98% of the human body is made up of water, the health of our water systems affects human health greatly. Many of these programmes have international components. Check out the following websites to see some of the work on these issues. Are similar programmes being replicated in India? What are the groups doing such work in India? See what you can discover.

  - Wetlands Protection (www.audubon.org/campaign/cleanWater2.html)
  - Long Island Sound (http://ny.audubon.org/lis.html)
  - Greenpeace (www.greenpeace.org/international/campaigns/oceans)
  - National Resources Defense Council (www.nrdc.org/water/default.asp)
  - Sierra Club (www.sierraclub.org/communities/cleanwater/default.aspx)

- **Waterkeeper Alliance:** Started by Robert F. Kennedy, this organization provides means and support to help citizens become watchdogs of their water (www.waterkeeper.org)
  - Information on water use around the world, water footprint calculator (www.waterpressures.org)

11.5   ENVIRONMENTAL HEALTH AND DISEASE

There are many toxins in our environment resulting from production and usage of manufactured consumer products. Many believe that modern diseases are directly a result of this lifestyle. All nations are faced with health issues due to toxins, which are worthy of research by students:

- Air pollution and diesel particulates are linked to asthma, cancer, cardiopulmonary ailments, and premature death.
- Pesticides are linked to cancer, reproductive problems, birth defects, lung and kidney damage, headaches, nervousness, nausea, cramps, and diarrhoea and they are a trigger for asthma.
- Lead poisoning leads to delays in normal physical and mental development, deficits in attention span, hearing loss, learning disabilities, stroke, kidney disease, and cancer.
- Mercury is linked to developmental problems in the brain, spinal cord, kidneys, lungs, and liver. It is present in mercury thermometers, fluorescent light bulbs, batteries, light switches, electronics, even dental fillings, and in many waterways due to improper disposal and release through manufacturing processes. Incinerators, coal burning power plants, and other industries release mercury in their air emissions. E-waste recycling industries in India can also be sources of mercury. Women of childbearing age and children under 15 are advised not to eat certain species of fish because they are contaminated from living in poisoned waterways.
  - Chlorine-based chemicals are linked to neurological diseases, hormonal and immune system disruptions, infertility, cancer, and birth defects. All are considered some of the worst toxins because their production, usage, and disposal are all sources of contamination. Dioxins are created when chlorine-based chemicals are produced, used, or burnt and remain in a person's body and in our water, air, soil, and food. These toxins do not break down and cause damages for years to come. Some chlorinated chemicals are as follows:
  - PCE/PERC (perchloroethylene solvent used in dry cleaning)

- PCBs (polychlorinated biphenyls, used in electrical insulators, discharged in our waterways, and now concentrated in animal flesh, milk, and eggs)
- PVC (polyvinyl chloride in vinyl, many toys, plastic wrap, and plastics)

- Modern processed/fast food diets (excessive fat, sugar, salt, preservatives, processed foods, additives, pesticide and herbicide residues, hormones, antibiotics, and bacteria) are linked with increasing rates of cancer, stroke, arteriosclerosis, heart disease, hypertension, diabetes, early puberty, premature ageing, and obesity. Additives and excessive amounts of chemicals and lack of nutrients in modern, processed, factory produced foods are associated with multiple health problems, including an epidemic of obesity in America and increasingly in the developing world as these nations adopt a Western diet.

- Cigarette smoking is linked to asthma, respiratory illnesses, birth defects, and cancer. Second-hand smoke can increase the risk of these diseases as well, especially in children. Advertising and sale of cigarettes to youth are special concerns.

- Drugs and alcohol are linked to birth defects, addiction, cirrhosis of liver, cancer, and premature death.

- Animals and toxins are also an area of study. Scientists who monitor populations of birds and other wildlife can use their studies to find threats to the health of humans and the overall environment. DDT was found responsible for the death of many birds before being banned in the 1960s, and lead sinkers, PCBs, pesticides, and herbicides are indicators today. Fish, birds, and other wildlife are more quickly affected by concentrations of pollutants in the environment and thus a lowering of populations of animal species can predict impending harm for humans. Since many toxins concentrate in fat, animals that we eat, such as fish, wild game, cattle, chicken, and hogs, and their milk and eggs often are sources of concentrated toxins when we eat them. To preserve biodiversity and our own health, curbing the use and production of toxins is essential.

- Environmental justice and health is another issue to research. Often impoverished communities and people with lower level of education find themselves living in areas and conditions where all the aforementioned environmental toxins are present from multiple and overlapping sources. Excessive siting of sources of pollutants (diesel bus depots, sewage treatment plants, industries, landfills and dumps, highways) and lack of education and information about the dangers of these toxins in such communities make problems much more difficult to solve. Poor citizens may even work in harmful unregulated industries and be exposed to more pollutants. Children, whose bodies are smaller and less capable of processing toxins, are more vulnerable.

Investigate information about toxins available at Natural Resources Defense Council (www.nrdc.org/health and www.nrdc.org/air/default. asp) and Greenpeace (www.greenpeace.org/usa/campaigns/toxics).

11.5.1   Environmental Health and Disease-related Activities

- Carbon dioxide (CO_2), a major greenhouse gas (GHG), is of the greatest concern because of the large amounts that are generated from the exhaust of gas and diesel engines, from power plants that burn coal or oil, and from the removal of CO_2 absorbing forests. Overall, plants remove CO_2 from the atmosphere and animals contribute it in the carbon cycle, but the other sources overwhelm the natural cycle. To understand a little more about the properties of CO_2, try the following to generate some: mix an acid such as vinegar with baking soda in an acid-resistant container (glass or plastic). As CO_2 is heavier than air, "pour" some of it onto a lit candle or even down a sloping trough with several candles arranged at intervals. If enough CO_2 is generated, the candles will go out as the invisible gas flows down the trough. Calcium carbonate (limestone) and hydrochloric acid (HCl) can also be used instead. Use proper safety precautions, particularly with HCl.

- Investigate some of the properties of CO_2. Compare its molecular weight with that of air, which is approximately 29. Look up some of the uses of CO_2, such as in fire extinguishers and as "dry ice". Write the balanced chemical equation for producing CO_2 from baking soda ($NaHCO_3$) and HCl.

- Audit the use of chemicals in institutions or home. Conduct surveys or examinations of all chemicals used and research ingredients and their effects on health. You can include foods and air and water quality testing. Lobby to change methods of cleaning and food sources.

- Test lung capacity. With the help of simple devices, students can measure and compare their lung capacity with that of asthmatic individuals.

- Map and study the sources of pollution in communities. Individuals can map locations of producers of toxins and pollutants in their neighbourhood, high traffic areas, and homes of severe asthmatics. Lobbying could be done to change some of these situations if appropriate.

11.6 WASTE MANAGEMENT

The more a nation industrializes, the more waste is generated, which cannot be decomposed naturally. This creates toxic landfills, health hazards, air and water pollution from incineration, unsafe disposal of other wastes, dangers to wildlife that ingests wastes, and excessive amounts of money spent on waste disposal.

Nations are using three basic concepts to encourage rethinking and reorganization of waste in the modern era: reduction in waste generation, recycling, and reuse. All overlap and are interconnected, but thinking about them separately gives a clear understanding of each.

1. **Waste reduction** involves finding ways to reduce the generation of trash by purchasing less, packaging less, and adopting other means. In this scenario, a conscious effort is made not to create waste.

2. **Recycling** is taking unwanted items and reprocessing them to make the same product a second time. Paper, metals, plastics, and glass are the most commonly recycled materials around the world. As more awareness grows, more and more products are increasingly being recycled: crayons, printer cartridges, and toothbrushes are some unique items being recycled. Students can investigate what products are recycled in their communities and around the world.

3. **Reuse** is utilizing again a product that might normally have been discarded or donating the product to someone else to be used again. Many places have thrift stores where unwanted items can be deposited to be refurbished or resold.

The following activities investigate waste management:

- Investigate the general waste management in your city. Find out how and where your trash goes in your community. What is your locality's waste management system? Is your waste dumped in landfills? Is it incinerated? Are there people who recycle your waste before or after general disposal? How do industries dispose of waste? What regulations are there, and are companies following them? What waste from industrialized nations is being dismantled in India and how? Gaining awareness is the first step. A simple project is to audit inventory, and weigh your trash.

- Composting is allowing plant waste to break down naturally and turn back to soil. This practice is ancient and necessary in rural/agricultural communities. Are any homes, local gardens, schools, or businesses composting in your community? It is estimated that about 60% of most urban waste could be composted, and it could considerably reduce the need for waste disposal. Investigate whether such a programme can be implemented or where you live. If you have the space, start and maintain a compost pile. As what is available in a household and what can be done vary widely across India, try to locate local sources of practical advice and information. Be sure that whatever is done is acceptable, both legally and by neighbourhood standards.

- Quantity-based user fee is the concept of having people pay for the amount of garbage they create. This is being implemented in many large cities around the world as a way to create awareness and responsibility for one's own waste. Is this happening in India?

- Extended producer responsibility is a new and growing concept that makes manufacturers responsible for their product and packaging from production to disposal. Currently, electronic companies are leading in this area, and many have take-back programmes, when you want to get rid of an electronic device such as a computer. This responsibility makes industries more reflective about how a product is produced to make sure its components can be recycled when disassembled. E-waste is shipped to India for dismantling, and so this is an important area to research to understand India's place in this cycle.

- Research and create strategies for waste reduction at home. Challenge friends and family to buy less. Use both sides of paper for printing and copying. Always consider how an item will be disposed of before purchasing it. Aim for zero waste and research groups promoting this philosophy. NGOs in India are leading the way towards zero waste management in cities and villages.

- Investigate if there are refurbishing centres, recycling plants, and demolition depots in your community. Could there be more for other products? Be sure to donate all and any items to these places if you no longer want them.

- Brainstorm ways in which simple products can be reused at home: jars for storage and paper for notepads. Cooking oil can be turned into biodiesel if there are facilities available. Research them.

11.7 FURTHER ACTIVITIES, PROJECTS, AND RESOURCES FOR GREEN AND SUSTAINABLE LIVING

11.7.1 Additional Activities

- Study chemicals in ordinary foods by researching ingredient labels of favourite brands and their health effects. Create a plan to modify diets.
- Create a green business and products source list or map for your community. This would include lists of where people can go to buy green, organic, or recycled products or participate in activities that are good for the environment.
- Survey and make green recommendations to schools/local businesses for how to be more environmentally responsible in a comprehensive way.
- Utilize and calculate one's ecological footprint (www.myfootprint. org). This is a way to measure the amount of land and water used by each of earth's 7 billion people to provide the resources needed to live and to accommodate the wastes they make. In the United States, the average ecological footprint is 31 acres per person, but in reality the world's landmass allows only 5 acres. This equals roughly 4 pounds of garbage per day and 2500 gallons of fossil fuels per year. Europe has a lower footprint, with less space and fewer resources; Europeans' footprint is about 15 acres.
- Since the 1970s when the first Earth Day was celebrated, many Americans have become increasingly aware of environmental threats to our planet. Some problems have been reversed; leaded fuel is no longer used, and DDT is now banned, as evidenced by the bald eagle's comeback in the United States. More laws are in place to protect the environment. Research the history of the implementation of such laws and find out which are being currently lobbied in your nation. Find ways to participate in the dialogue.
- Assess what you consider the major concerns involving the environment. There are various ways to approach this evaluation:
  - Each person first writes what he/she thinks are the concerns and then ranks his/her items from the most important to the least.

- With a small group of four or five individuals, record the top three concerns from each individual. Discuss the list and combine those that are essentially similar; then have each team member privately rank the items in order of importance.
- Combine the private rankings into a group list and present that list to the rest of the class with some examples or other persuasion as to why the group chose as it did.
- As there are so many possible concerns, limiting the list in some way is typically more productive. The limitations could be by area, be it state, country, region, or world; by topic such as water related; by specific issue; or by time, recent versus long term.

- Make a picture series, slide show, or video/movie of some environmental topic. For example, follow a trash pick-up, home, or school to disposal, or make photos of pollution or a degraded area.
- Compile environmental news items using whatever sources are available: newspapers, magazines, TV, online, or even word of mouth. Discuss what you have found, including opinion, and analyse whether the material takes a specific point of view.
- Select an environmental problem at home, neighbourhood, state, country, or the world. Write a questionnaire about the problem and take the questions to various people to see how aware they are about the issue and what opinions they have. Be careful that the questions are neutral even if you have a firm opinion about the issue. If possible, compare perceptions obtained from the questionnaire to the available facts.
- Some in the business and scientific world have devoted themselves in promoting and creating alternative non-fossil fuel energy sources, energy efficient appliances, and organically grown and processed foods, cosmetics, cleaning supplies, and other "green" consumer products and services. For almost any product you wish to purchase today, there is a more environmentally friendly version—one that has been made without the use of chemicals, made from recycled or reused materials, or harvested and made in such a way that the environment was not harmed at any step. Investigate companies doing this in India and find out where you can buy such products. Some may have been around for a long time. Try to support these fair trade, cottage industries so that they do not go out of business because of the availability of new products.

- Greenwashing is an issue to be concerned with. Many businesses originally not part of the effort are beginning to incorporate ethics and environmentally and socially responsible practices in production due to consumer pressure. At times it is difficult to determine if a business is seriously committed to sustainable development or if it is just good public relations. Compare and contrast various businesses and their practices to see if you can find examples of greenwashing. Two international and large companies that have been criticized, but that seem to be making real strides are Home Depot and Walmart. Walmart is working with an environmental organization in the United States called the Environmental Defense Fund.

- Many other organizations have been formed to promote consumer awareness about these alternatives and to advise on many other steps individuals can take to live each day consciously. They work continuously to make people aware of the impact their actions, shopping habits, and lifestyle have on the environment and also other people in the world. The word "sustainability" is often used to define this concept of living in such a way that we humans do not exploit others or overtax the natural systems of the planet to the point that it becomes uninhabitable. Research the work of these organizations and get involved with their efforts by teaching friends and family about these concepts.

- The growing movement of citizens, business owners, and leaders around the world who are trying to create a just and sustainable world is inspirational. Research these leaders and get involved.

- Interview someone in the environmental field who lives or works in your area.

- Research sustainable energy developments. The United States has about 3% of the world's oil but consumes 25% of the world's oil. Energy dependency on fossil fuels creates pollution, excessive amounts of CO_2 and global warming, and political and economic crises. There are blackouts, constant political tensions in the Middle East, and debate over whether to allow drilling in national parks such as the Arctic National Wildlife Refuge in the United States. Efficiency and alternative sources must be developed. Solar power, wind power, geothermal power, and hydropower are being increasingly used for electricity. Electric, natural gas, and biofuel (cooking oil, hemp, mustard seed oils) powered cars are being developed and researched. See what you can find out about alternative energy use in your community.

- Green buildings are becoming more and more popular. Engineers and architects design buildings to be energy efficient, use fluorescent lighting, low-flush toilets, light sensors, recycled or non-toxic building materials, and other techniques to create the most environmentally friendly structures possible. What are the green building initiatives in India? What are organizations, such as The Energy and Resources Institute and the Department of Science and Technology, doing in this area?

- Population growth also contributes to current environmental stresses. If each person has one child (two per couple), zero growth can be achieved. Research this issue and see how well India is doing and how it compares to other nations.

- Investigate social investment in phone and credit card companies. Certain banks and mutual funds now screen companies that do business in environmentally unfriendly or other unfair ways. Consumers can also buy stock in companies they most want to change and submit shareholder resolutions to push for them. Some phone and credit card companies now offer ways to make automatic donations to non-profits working for a better world.

- Analyse the sorts of activism in India for environmental issues, either as specifics such as the protests related to the Bhopal tragedy or in general. This includes demonstrations and marches, letter campaigns, petitions, attending meetings, discussion with regulators and elected officials, involving the press, or even hunger strikes. Activism has a definite tradition in India.

- There are special concerns in very poor communities as far as creating sustainable lifestyles and green citizens. Consider the following:

  - The poorest people around the world will suffer the most as resources are depleted, habitats destroyed, and climates change. They will not have the means to afford scarce resources or, if employed as farmers (as many are), make a living if there is no rain, excessive flooding, and so on.

  - Often the impoverished have very little information or education regarding these issues, which compounds the problems.

  - Sustainably produced products or services are often costlier and unaffordable.

- Compare the energy efficiency of light bulbs. Many environmental decisions involve costs, but costs are often difficult to determine, particularly when the benefits are diffused and the costs are borne by a few. An easy example to grasp is comparing light bulbs: one with a longer life and energy efficiency but expensive versus one with a shorter life but cheap. Obtain information about the costs of the bulbs. Usually the packaging will give the needed details; what is needed is the power of each bulb, expressed in watts.

The energy efficient bulb will typically say what it compares to for brightness; for instance, an energy efficient 16 W CFL bulb may be as bright as a 60 W regular incandescent bulb. You can also compare LED bulbs.

Determine how much it will cost to power each bulb for a period of time using electricity rates. Add up the overall costs of the bulbs themselves plus the energy to run and the lifespan by using packaging or web research and make a table to chart your information (Figure 11.1). Which bulbs are better to use?

Figure 11.1   Energy efficient CFL bulb versus incandescent bulb versus LED bulb

11.7.2   Extended Projects

11.7.2.1   *Solar Cooker Project*

In some parts of the world, cooking is done over smoky wood fires. These fires are often made with hard-to-obtain fuels that not only take time to gather but also lead to deforestation. Even when fuel is abundant, not sticks of wood gathered with great effort from a dwindling resource, the smoke irritates the eyes and lungs of young and old. At the same time, most of the same places, for many months on end, have the advantage of predictable sunshine, which could be used for cooking.

To understand this potential, build a solar cooker (Figure 11.2). As specific plans might require unavailable materials, consider the cooker simply as an enclosed space pointed at the sun with a baking dish inside. The sunlight must be concentrated by reflection and the heat retained.

Figure 11.2 Solar cooker

A simple, inexpensive cooker could be a cardboard box with the flaps painted white, a transparent cover, and some insulation. A sturdy rack for the dish is also essential. Use your imagination and what is available to come up with variations: Nest one box inside another for better insulation, replace the cardboard with a wooden box, and use aluminium foil for the reflecting surfaces and pieces of old rug for insulation. Note that because there is no flame, there is no danger of burning the food and the pot does not have to be watched closely. However, the cooking time is typically several times greater than that of using an open fire. Set a goal for your cooker. For example,

- Heat a litre of water to a certain temperature, even to boiling.
- Make tea.
- Cook some rice; compare the time needed to cooking over a flame.
- Cook your favourite recipe for dal. Compare the taste to the usual method.
- Bake some roti. How does the taste and texture compare?

Evaluate the following questions:

- If your cooker is made of cardboard, would it catch fire?
- What kind of dish is best, of what material, thickness, and colour?
- How necessary is it to track the sun carefully? How much sun is adequate?
- Does the food cooked in a solar cooker taste different from the food cooked traditionally?
- How much does quantity matter? Will one cup of rice take twice as long to cook as half cup?

- Would the cooker heat contaminated water enough to make it safe?
- In different parts of India, which are the best times of year to use a solar cooker? Are there times when it would not work at all? How much sunlight is enough?
- A solar cooker would be useful in rural areas. What problems are there to getting greater use in those areas?
- Could a solar cooker be useful in urban areas?
- How much fuel can be saved in a week, month, or year? How much saving will that bring of time and expense?
- All regular fuels produce CO_2, a GHG. How much would a solar cooker reduce GHGs?

11.7.2.2 Carbon sequestration by trees

- First select a tree in the area where you live.
- Measure the diameter and height of the tree. Exact values are not essential; a good estimate is the amount of increase in diameter of the trunk of the tree in a year.
- Using those numbers, calculate the volume of wood added to the tree. Assume that the trunk of the tree is a straight cylinder and calculate the volume before and after the increase in diameter. Subtracting will give the increase in volume. Make sure the units of the measurements agree.
- Convert the volume to mass. The mass of water is 1 kg per 1000 cm^3 (0.001 m^3). Use 8 as the specific gravity of wood, that is, wood is eight-tenths as dense as water. This is approximate. While almost all wood is less dense than water, different woods have different densities.
- Calculate the mass of dry wood gained by the tree. The moisture content of wood is about 0.5. This is also an approximation.
- Figure the weight of carbon taken up in the mass you obtained. Assume that wood has the formula $C_5H_{10}O_5$, and figure the fraction of C in the formula. The major component of wood is cellulose, which is made up of many units of $C_5H_{10}O_5$.
- Knowing that CO_2 is 3.7 times as massive as carbon (44 amu for CO_2 and 12 for carbon), figure the mass of CO_2 that is removed from the atmosphere by the one tree. The uptake of carbon (as CO_2) by trees and other plants is carbon sequestration.

11.7.2.3   Carbon dioxide generated from a vehicle

Almost no one realizes the amount of CO_2 produced by a motor vehicle (Figure 11.3). What would you guess for a year, perhaps one-tenth of the weight of the vehicle? Maybe more, maybe less? You will be able to estimate that amount with the following calculation:

Petrol is a complex mixture of hydrocarbons—compounds made up of mostly hydrogen and carbon—although some petrol components may also contain oxygen. Regardless of when they burn, most of what is produced is simply CO_2 and water. Other products such as carbon monoxide do not make up much of the weight of the exhaust, and hence they can be neglected for the calculation. They cannot be neglected when pollution is considered, as even small amounts are dangerous.

For the calculation, assume that petrol is made up of only isooctane (2,2,4-trimethyl pentane). Isooctane is one of the hydrocarbons used as a standard for rating fuels, the octane number.

(1) Knowing that octane (C_8H_{18}) burns by combining with oxygen (O_2) from the air to produce CO_2 and H_2O, write a balanced equation for the process.

(2) Calculate the weight of 1 mol of octane, and then calculate the weight of CO_2 produced by burning that amount of octane.

(3) Calculate the number of litres of petrol burnt in a year by a motor vehicle using these assumptions: The vehicle is driven 15,000 km in the year and uses 1 L petrol for every 20 km.

(4) Calculate the weight of the octane burnt in the year using the fact that a litre weighs 0.69 kg. Then using the weight ratio from step 2, calculate the weight of the CO_2 produced.

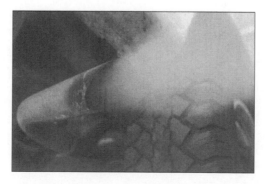

Figure 11.3   Exhaust gases from a motor vehicle

(5) Compare the weight of CO_2 produced to the weight of the vehicle, around 1 tonne.

(6) Finally, give your reaction to what you have calculated. Is the weight of CO_2 produced more or less than you would have thought? Is the calculation realistic or are there changes you would make in the assumption? Because CO_2 has been implicated in global warming, do you think one vehicle driven for 1 year makes a noticeable contribution to global warming?

11.7.3 Other Resources for Sustainability and Green Living

11.7.3.1 *General*

- **Care 2 Make a Difference** (www.care2.com and www.ecologyfund. com): Extensive list of websites with links to shopping for green products as well as information on multiple environmental issues.

- **Co-op America** (www.greenamericatoday.org): National membership organization actively promoting sustainability through public education initiatives and research on social investing, renewable energy, deforestation, sweatshops, and many other issues. Publishes national green pages and hosts multiple websites.

- **Dump Scorecard** (www.scorecard.org): Site that allows you to enter your zip code and generate a list of companies and the poundage of toxic waste they have contributed to the environment.

- **E magazine** (www.emagazine.com): Educational magazine with stories on the most current environmental issues and initiatives.

- **Earth's 911** (www.1800CLEANUP.org): Extensive information, zip code by zip code of how you can live in an environmentally friendly way.

- **New Dream** (www.newdream.org): Organization promotes sustainable living, less consumption, and care for the environment to create better paying jobs, more satisfaction, more money, and more free time for everyone.

- **National Oceanographic and Atmospheric Administration** (www. noaa.gov): Current information about weather, climate changes, and natural phenomena in the United States.

- **Responsible Shopper** (www.responsibleshopper.org): Links to places to buy sustainably produced products.

- **Teaching about Climate Change** (www.greenteacher.com): Lesson plans for teacher, but interesting for all.

- **World Wildlife Fund** (www.worldwildlife.org):   Has various ways to get involved.
- **Global Green USA** (www.globalgreen.org):   International network working to green the world.
- **The Apollo Alliance** (www.apolloalliance.org):   Coalition to build sustainable energy networks and green jobs.

11.7.3.2 Food

- **Just Food** (www.justfood.org):   Promotes the creation and just and healthy food for all in New York City. Links upstate organic farmers with city residents through CSA. Residents purchase shares for a season and pick up fresh vegetables weekly at designated locations in New York City. Similar projects exist around the world or can be started by those interested.
- **Farm Animal Reform Movement** (www.farmusa.org/):   Promotes a vegetarian and/or vegan diet to reduce suffering of factory farmed animals, curb pollution from such practices, and stem the rates of cancer, stroke, obesity, and other diseases in America. Sponsors Great American Meat-Out on 20 March.
- **Heifer Project** (www.heifer.org):   Fundraising to purchase animals, trees, bees, and other wildlife to send to developing countries. Families become owners of livestock and have new sources of income.
- Activities to do at dinner parties to promote thought about our food sources (www.earthdinner.org)
- Helps people find healthier eating habits: www.sustainabletable. org
- **Slow Food Movement** (www.slowfood.com):   Provides many reasons and efforts to slow down the trend towards fast, unhealthy food.

11.7.3.3 Energy

- **Alliance to Save Energy** (www.ase.org):   Provides educational materials on how to conserve energy to save money and the planet. Has created a manual called *Green Schools*, a comprehensive programme designed for K-12 schools that creates energy awareness, enhances experiential learning, and saves money of schools on energy costs.

- **Get Energy Smart** (www.getenergysmart.org): New York State's website to teach New Yorkers how to make their homes more energy efficient.
- **Solar Cooking** (www.solarcooking.org): Teaches people how to make solar cookers. Especially important in areas where there is no electricity.

11.7.3.4 Consumer products/services

- **Alternative Gifts** (www.altgifts.org): Sells gifts that are not given to the receiver, but rather are donations to non-profits and development projects for communities in need. Reduces consumption, and steers spending and gift-giving to those who really need it.
- **Amazing Recycled Products** (www.amazingrecycledproducts.com): Extensive catalogue of products made from recycled materials.
- **Eco-mall** (www.ecomall.com): Source for shopping for sustainably made products.
- **Red Jelly Fish** (www.redjellyfish.com): Source for shopping for sustainably made products.
- **Shop for Change** (http://shopforchange.in): Source for shopping for sustainably made and fair trade products in India.

11.8 CONCLUSION

Humans are utilizing natural resources and altering the climate and natural systems at a rate and in ways that may eventually make the earth uninhabitable. Wildlife species are disappearing, green and natural spaces are decreasing, and the quality of air and water does not meet established standards. Water is scarce, health problems of humans and wildlife can be linked to toxins in our environment, and we have overwhelming amounts of garbage. Global warming and related climate change due to the usage of fossil fuels, melting of polar icecaps, and rising sea levels have been documented. Lack of resources, degradation of the environment, and violations of human rights due to unjust and unsustainable business practices can create tensions that can escalate into wars.

This increased knowledge of the impact of human lifestyles and consumer choices forces everyone in the modern world to consider their actions and impact on the environment. While no substitute for political and industrial-level decisions, taking responsibility individually is another way to work towards solving the world's environmental problems. "Green citizens" take the concept further by not only living sustainably and reducing their footprint, but also being active advocates through voting, community participation, investing, and active letter writing, lobbying, boycotting, or other involvement in promoting environmental awareness in the entire lifestyle.

Colleges should work to create comprehensive environmental education programmes that include the variety of topics discussed in this chapter. This field of education is constantly and rapidly evolving. More and more resources are available to assist educators or students wishing to develop a full curriculum in their school. The ideas and resources listed provide a sample of existing initiatives.

12

Indian and International Case Studies

12.1 INTRODUCTION

Modern society is dependent on a large number of chemicals and plastics, which are pervasive in our life. From fertilizers and pesticides to cell phones, computers, and houses, we use chemicals and plastics everywhere. The improper disposal of waste from chemical and plastic industries has destroyed land and groundwater resources in many parts of the world.

The impact of contaminated sites on the lives of poor people is considerable, and, hence, prevention of such contamination is the key to managing hazardous waste. Proper recycling through treatment and disposal using engineered facilities is critical in managing hazardous waste sites. Good governance is important to enforce the "polluter pays" principle, and a strong court system is also necessary.

12.2 LOVE CANAL: BIRTH OF HAZARDOUS WASTE SITE CLEAN-UPS

In the United States, the US Congress passed the Comprehensive Environmental Response, Compensation, and Liability Act in 1980 to investigate hazardous waste sites, find out the parties responsible for the sites, file lawsuits to recover funds for the clean-up of the sites, and finally clean up the sites to prevent further environmental damage. A tax was imposed on oil and chemical industries, and a fund called the "Superfund" was set up to clean sites where a responsible party could not be found to do the clean-up.

Love Canal is a neighbourhood located in the City of Niagara Falls, New York. It was started in the early 1900s to provide clean housing for the residents employed in the industry located in the vicinity. Cheap electricity from Niagara Falls was the impetus for building the neighbourhood (Figure 12.1). However, by 1978, Love Canal became a national scandal when it was discovered that the neighbourhood was built over tonnes of toxic waste that was improperly disposed of by the previous owner of the land (Bryan 2004).

Figure 12.1   Location of Love Canal

Hooker Chemicals had disposed of hazardous chemical wastes in the Love Canal area. When they closed the disposal site in the 1960s, they capped the site with clay and soil. They sold the land to the City of Niagara Falls for a dollar, stating that hazardous chemicals were disposed on the property. However, no record of the existence of the chemical waste dump was placed in the property deed, and, thus, was not revealed to the developers of the land. As homes were built on that land, the chemicals leaked into the soil and groundwater. Symptoms of leukaemia were witnessed as polychlorinated biphenyls (PCBs), benzene, and dioxin had leaked into the soil and water. In 1978, an investigation conducted by the New York State Department of Environmental Protection and the US Environmental Protection Agency (EPA) revealed that benzene and dioxin levels in the soil were 100 times the normal level in the area (Bryan 2004).

Occidental Petroleum had purchased Hooker Chemicals, and more than 640 lawsuits were filed against Occidental to recover $14 billion in damages. In 1978, 239 families were evacuated from Love Canal, and during 1979 to 1980, 900 families were evacuated by the government to avoid further impacts on the population. The government spent $250 million to clean up the site, and by 1999, the site was cleaned up using geomembranes and a clean soil cover. Several families moved back to the site after the clean-up.

However, portions of the site remained permanently fenced. In 1995, Occidental settled the claims and $129 million worth of damages were paid out to the families.

Of the 1200 sites listed on the National Priorities List assembled by the USEPA, hundreds have been cleaned up and several are in the process of being cleaned up. It is extremely expensive to clean the soil and water contamination from hazardous waste sites, and billions of dollars have been spent by the United States. Some of it has been paid by the US taxpayers, but most of it has been collected from the responsible parties through the "polluter pays" principle. In addition, a tax on oil and chemical companies was collected by the government to pay for the clean-ups.

12.3 ALANG SHIPYARD

"Every day one ship, every day one dead." With that macabre phrase, one gets a rough feeling of the Alang Shipyard in Bhavnagar, Gujarat. The shipyard on the coast of the Gulf of Khambat (Cambay) (Figure 12.2) is both a major employer and a cause for environmental concern. The work consists of shipbreaking, scrapping of obsolete ships, accounting for around 50% of the world's total ship recycling. The shipyard provides jobs to tens of thousands of workers belonging to Odisha, Uttar Pradesh, and Bihar. They live in substandard lodging near the work site and send money back to their villages.

Several months are needed for breaking up a large ship. Before Alang, shipbreaking was done in regular shipyards with docks and cranes, in Europe, North America, Japan, Korea, and Taiwan. As the same result could be achieved at Alang with a much lower cost of labour and fewer environmental and safety restrictions, ship owners started selling their obsolete ships to Indian shipbreakers, thus gaining cash for themselves and providing opportunity to the shipbreakers. The 10 km coastline has more than 170 shipbreaking plots, not all constantly active. As the business is cyclical, activity moves somewhat with the global economy. Surprisingly, a good time for shipbreaking is when the global economy

Figure 12.2 Alang Shipyard

is weak since ships not in use are a cost to the owner. In other words, a recession in the world can mean business for Alang.

The basic scrapping process is straight forward. Ships are run aground on the beach at high tide and some parts are removed as the superstructure is cut away. As the tide goes out or the hull lightens and is winched farther up the beach, the process continues until nothing is left.

This description greatly oversimplifies the process, neglecting the overall complexity—the equipment needed, the skill, coordination and safety awareness, the destination of the parts and metal, and the junking of unusable material. Although attention is often drawn to the fixtures salvaged, the iron of the hull is what is most valuable when sold and recast. As there will be junk, the earlier phrase "nothing is left" is not that accurate and even with the best practices, there will be spillage and waste during the scrapping process, contaminating the sand and water.

In a ship, hazardous materials of different kinds can be found such as, fuel oil still coating the tanks and lubricants of different forms, PCBs as electrical insulators, metals of varying sorts, and asbestos insulation. And there are risks of the job itself beyond those materials. A beached ship has the height of a multi-storey building—falls can be fatal, loosened pieces can fall on those below, and burns and explosions threaten. After an explosion killed more than a dozen workers, standards were tightened for clearing flammables before the cutting torches went to work. With the nearest full-scale hospital in Bhavnagar, 50 km away, prompt treatment of accident victims is limited.

Because of the pollution and workplace conditions, the shipyard was criticized by environmental groups, particularly by Greenpeace. In 2002, Greenpeace members, posing as buyers, visited the Alang Shipyard to gain information and photos "to support its protests" of the "dangerous conditions" for the workers and the "toxic waste produced through ship-breaking". Since then access to the shipyard is greatly restricted.

By contrast, besides employment for thousands, the Gujarat Maritime Board (GMB) emphasizes the other economic advantages that come with the shipyard. While the road to the shipyard, lined with ship fittings and furniture, is the most obvious result to those nearby, the Gujarat Rolling Mills producing steel reinforcing bars (rebar) and oxygen refillers are dependent on the yard for material or business. Additionally, the GMB assesses a fee by tonnes of ship displacement.

Some argue that workers are not forced to come to Alang and that they would earn less in their home state, a point that is consistent with the local objection to the French aircraft carrier *Clemenceau* being forced back to France. Another argument is that while Alang does 50% of the world's shipbreaking business, too many restrictions or too much concern for workers and the environment would send the business to Indonesia, Pakistan, Bangladesh, and elsewhere, and the beaches of Alang would be left with nothing but the high tides that enabled the business in the first place.

Thousands of small and large ships have been broken at Alang since shipbreaking started in 1983; however, three have drawn the most attention for a mix of reasons—their size and toxic contents, legal issues, and their international connections.

According to estimates, *Clemenceau* still had 45 tonnes of asbestos on board when it arrived off Alang. After international protests, particularly about the risk of the asbestos aboard, Jacque Chirac, then the president of France, recalled *Clemenceau* in 2006 to remove asbestos in a French port or elsewhere. Asbestos was a favoured material for the insulation of steam pipes, serving as a fire barrier, and other uses. Unrecognized or avoided for some time were the details of the major health concern when asbestos dust is inhaled or absorbed. Asbestos fibres lodged in the lungs damage them like tiny glass needles or bits of sand, developing a condition called asbestosis, which can develop in just a few years. Asbestos also causes a form of cancer, called mesothelioma, which has a longer time of development. Smoking multiples the chance of asbestosis and lung cancer, an effect called synergy, where the combined effect of two things, in this case smoking and exposure to asbestos, is greater than twice the effect of either one. Asbestos is also used in consumer applications such as brake linings, floor and roof tiles, and wall board, but since the exposure at home is less than that at the shipyard or in construction, the dangers are less.

Next was the *Blue Lady* of 46,000 tonnes, launched in 1960 as SS France as a passenger liner, later owned by Norwegian Cruise Lines and used as a cruise ship, the SS Norway.

Some sources said the ship contained 900 tonnes of toxic materials, including asbestos and the radioactive isotope americium 241. The ship was beached at Alang in August 2006. Various injunctions wound their way through the courts, but the Supreme Court ruled in September 2007 that salvage could begin and shortly after it did.

Finally came the SS Oceanic, originally from the United States and launched as the Independence in 1950, but owned by Norwegian Cruise Lines before they sold it in 2008. It was beached at Alang in April 2008 with some environmental objection, although less than that for the *Clemenceau* and the *Blue Lady*.

An estimated 250 tonnes of asbestos was reported to be present in the ship, along with around 210 tonnes of PCBs. Since PCBs are non-flammable oil-like liquid, they make good insulators for transformers and other electrical applications. Their drawback is that they are toxic to both humans and wildlife.

These three ships received the most attention, overshadowing the hazards posed by thousands of other ships.

Started only a year before the Bhopal accident and three years before Chernobyl, the Alang Shipyard, though scarred with its own human and environmental problems, has persisted and continues as an economic endeavour. With the passage of the Basel Convention by the United Nations, the transport of hazardous wastes has been banned from one country to another. It is hoped that monitoring of these activities and implementation of the Basel Convention will prevent another Alang happening anywhere in the world.

12.4   BHOPAL GAS TRAGEDY

Of the myriad environmental episodes, the Bhopal gas leak is well known in the country and world. Even now, decades later, the case is still evolving with charges and countercharges, legal and humanitarian efforts, and demonstrations and memorials. The facts are indisputable—thousands of people died from a massive chemical leak from the Union Carbide India plant in Bhopal, Madhya Pradesh, starting in the early hours of 3 December 1984 with thousands more injured or affected permanently. The plant produced the pesticide carbaryl (Sevin) using methyl isocyanate (MIC) as an intermediate, and it was the accidental release of MIC that caused the deaths. Chemical reaction for the production of carbaryl using methyl isocyanate is given below:

1-Naphthol   + $H_3C-N=C=O$ $\xrightarrow{\text{catalyst}}$ Carbaryl

MIC

Water somehow got into an MIC tank, causing the MIC to heat violently, turn into gas, and escape from the plant. As it was heavier than air, it descended on the victims. The death toll was greatest among the poor living near the plant. Insufficient or improper function of safety features—some were even shut down—failed to contain the hazard at the plant. Inoperative or insufficient warning systems failed to alert people inside and outside the plant until it was too late.

Inadequate training, understaffing, poor design, and inadequate safety systems all played a role in this tragedy. As the plant was not active, standby conditions resulted in overlooking the potential for catastrophe, even though MIC needed to be kept near zero degree. Holding any quantity of MIC was a known and serious violation of safety practice. With production operations shut down, inaction of emergency systems was not recognized as a concern. At the plant itself, water might have got into the MIC tank by operator error while cleaning other reactor vessels, or corroded pipes or valves might have failed. The company even suggested that an employee might have sabotaged the operation, a dubious claim. Even sabotage would not have caused the many deaths if large quantities of MIC were not stored when the plant was not operating and safety and warning systems had been robust.

Outside the plant, night-time slowed the recognition of the danger and confused those trying to escape since people could not judge the source of the hazard or the safest evacuation route. All were hindered by the general confusion. Propelled by a light breeze, the heavier-than-air MIC spread along the ground into the nearby makeshift residential areas, sometimes killing children but sparing their taller parents. In one area, almost no one was spared, adults and children alike suffered a gagging death.

The exact amount of MIC released is uncertain. It was evidently many tonnes, perhaps as many as 45. Even a small inhalation is fatal. Two other tanks of MIC still at the plant, containing smaller amounts, were later converted to useable pesticide to avoid a subsequent disaster. The amount and toxicity of the chemicals still at the plant site, which was never properly cleaned up after it was abandoned following the accident, can only be estimated. Although less directly toxic than the MIC, the amount may be many times as much as the MIC. The abandoned plant remains a chemical wasteland, poisoning land and water (Figure 12.3).

The death toll remains uncertain as well. The official immediate death toll was 2259. The Government of Madhya Pradesh confirmed a total of 3787 deaths related to the gas release. A government affidavit in 2006 stated that the leak caused 558,125 injuries, including 38,478 temporary partial injuries and approximately 3900 severely and permanently disabling injuries. Others estimate that 8000 died within two weeks, and another 8000 or more have since died from gas-related diseases. Figure 12.4 gives the extent of MIC spread in Bhopal.

Figure 12.3 Abandoned Union Carbide plant in Bhopal

Even after more than 30 years, the effects are still afresh in the minds of survivors and others; there is illness and more is expected (Figure 12.5). Groundwater is contaminated. Payments, whether from the settlement agreement or the government, may not be reaching to all of those most afflicted. The plant sits untended, partially stripped

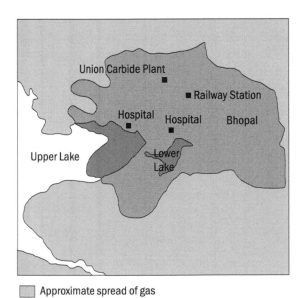

☐ Approximate spread of gas

Figure 12.4 Extent of MIC spread in Bhopal

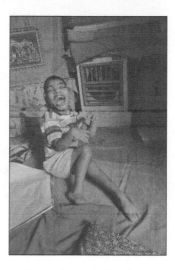

Figure 12.5 Victim of Bhopal gas tragedy

but a source of continuing contamination. Neither Dow Chemical, the purchaser of Union Carbide, nor the state and national governments are taking sufficient action to make things right.

When the plant was being built and first operated, equipment was state-of-the-art, proper procedures were meticulously described and observed, and training was extensive. Everyone was enthusiastic about the plant, the Indian Government for the technology that would help the country's crops, Union Carbide for the addition to their business in the country, and, not least, the residents of Bhopal for the opportunity to work for decent wages. As time went on, slippage began.

To detail all aspects of the tragedy and its aftereffects would fill volumes, in fact has filled volumes. Details though must not obscure the human loss from the disaster, nor overlook that the plant was built with careful oversight and the best of intentions. The erosion of those intentions and the slipping of oversight led inevitably to the catastrophe.

12.5 CHERNOBYL NUCLEAR ACCIDENT, UKRAINE

Chernobyl is one of the worst nuclear accidents. Chernobyl is upstream from Kiev, the capital of Ukraine. The Chernobyl disaster occurred on 26 April 1986 at the Chernobyl Nuclear Power Plant in then officially the Ukranian SSR (Figure 12.6).

No announcement was made to the world about the accident on that date and not until Soviet officials were forced to acknowledge something had happened when radiation was detected in Sweden about

Figure 12.6 Chernobyl Nuclear Power Plant

36 h later, brought there by the prevailing winds. The amount of release was massive, far greater than the bombs that destroyed Hiroshima and Nagasaki in World War II. Although without the blast effect of those bombs, the chemical fire at the damaged reactor continued to spew lethal amounts of radiation for at least several days, producing fallout that spread first across Ukraine and Belarus, then Western Europe, and then the world. As much as 60% of the total fell in Belarus, just north of Ukraine, both then part of the Soviet Union, but the radiation, carried by the shifting patterns of the wind and deposited by the rain, came to earth in patchy fashion.

A bomb is designed to release its energy within seconds, and it is the blast that does most of the damage to people and property, but those who survive the blast will receive enough radiation to cause illness, perhaps death, within days or weeks. Surviving that, in the longer term, there remains the danger of radiation-induced cancer.

Like a bomb, a nuclear reactor will have a critical mass of fissionable material, usually uranium-235. In contrast to a bomb, a nuclear reactor will have more nuclear material and must produce energy steadily, predictably, and safely. To achieve those goals, the neutrons that initiate the reaction in the uranium must be moderated and controlled. Lighter elements are the best moderators; hydrogen and oxygen in water and graphite, a form of carbon, are all suitable. Graphite is most commonly known as the "lead" in pencils, and graphite as the moderator in the Chernobyl Reactor was a key contributor to the magnitude of the disaster. Like a pile of burning coal, also largely carbon, when a fire starts, it is difficult to extinguish.

On 25 and 26 April 1986, a test was being run on the operating Reactor IV, which was near the end of a fuel cycle, hence operating at lower efficiency. Although those conditions were known to be dangerous by the designers and the military, but perhaps not the operators, safety systems had been turned off as part of the test. Somehow, through procedural error, rods in the reactor core, which should have been left in place, were removed. Early in the morning of 26 April 1986, without enough rods to control the extent of the nuclear reaction, overheating took place.

Triggered by those missteps, a cascade of failures combined to produce the catastrophe that is Chernobyl. Operators did not understand what was occurring. Control rods could not be reinserted and the overheating accelerated, causing first a steam explosion and then a hydrogen explosion, which toppled the heavy lid and released huge amounts of radiation in the air. This intensified the fire in the graphite with continuing release of radiation and started secondary fires. Fighting any large fire is dangerous, with smoke and fumes, carbon monoxide, and heat, but those factors in the graphite fire at Chernobyl, coupled with the high levels of radiation, converted the hazardous to devastating results.

Ground approach was deadly, and after several hours, the only thing that could be done was drop sand and other materials from helicopters to try to smother the blaze. Even so the graphite and contents of Reactor IV continued to burn until 5 May 1986, with a reoccurrence in mid-May. Even those brief flyovers were enough to expose the pilots and firefighters, the heroes of the response, to dangerous levels of radiation from the smoke and unmoderated nuclear reaction. Many of the operators at the time and some of the firemen died soon after of acute radiation sickness. The exact death toll is officially given as 57 by a 2005 international quorum—mostly plant operators and firemen and also some children from thyroid cancer. But the slow evacuation of the nearby town of Pripyat exposed many to radiation unnecessarily.

Although any accident is unexpected, precautions can reduce the risks to people and the environment—proper design and safety systems to reduce the chance of an accident, operator training, containment of the toxic material, and warning and evacuation plans. All these were badly neglected in Chernobyl, in callous disregard for the safety of thousands, made worse by operator error, confusion about what was occurring, and official mishandling.

Eventually, even with high levels of residual radiation as the fire died or was smothered, approach could be made to allow encasing the remains with concrete. Today, decades later, the wreck of Reactor IV is still in that sarcophagus with its stability uncertain and the evacuation zone still closed to permanent habitation in a circle with a 30 km radius. Towns were evacuated and remain so. Chernobyl, of course, and Pripyat, a town of 50,000 at the time, were both bigger and closer to ground zero, and those residents were part of the larger evacuation estimated at more than 300,000. Pripyat is 3 km from ground zero, built to house the workers and their families, remains a ghost town, unoccupied and decaying, visited only by officials and daylong tourist tours run from Kiev.

About 3–4 days after the initial failure, more was learned about the incident in Europe, where there was widespread concern, panic, and fear. Vegetables, especially the leafy ones, and fruits were avoided, both commercially produced and from home gardens. The apprehension was not based on testing of the produce for radiation, just the fear of the unknown. Certainly the undetectable nature of radiation without suitable equipment added worries and uncertainty, even mystery, accentuated by the reluctance of the Soviet Government to release information fully in a timely fashion. The available numbers for the death toll range from the official figure of less than 100 to thousands, even tens of thousands depending on the source. Most of the disagreement comes from trying to estimate the premature deaths accruing from the aftereffects of radiation, mutations and birth defects, anaemia, and cancers, from the acute immediate effects to the lingering, and some say the worst effect is psychological stress.

Economic repercussions are also only estimates, not just the loss of the reactors but of the electricity they would have been producing in the decades since, the expense of government services to the affected, the waste of crops across much of Europe and the continued ban on some food sources, the abandonment of Pripyat and other buildings in the 30 km zone, and the cost of building a replacement city, Slavutych. Even inexact, any estimate would be billions or hundreds of billions, likely among the most costly disaster in history. Effects of the accident are not known completely and cannot be, but they will continue to be debated. For one, cancer from radiation can take years or even decades to develop and even then misses some and strikes others. Still rainfall on the reactor site and its surroundings will drain into the Pripyat River leading to the Kiev Reservoir and continue to Kiev, the largest population centre in Ukraine. The radioactive run-off will continue downstream, the radioactive dust will continue to blow, and the evacuation zone will remain a wasteland, visited only by humans for a restricted time and inhabited by the wildlife of the fields and forests, unaware of the radiation that surrounds them or the accident that came before them.

12.6 NARMADA DAM CONTROVERSY

Part of the Narmada Valley Project, a large hydraulic engineering project involving the construction of a series of large irrigation and hydroelectric multi-purpose dams on the Narmada River, the Sardar Sarovar Dam is a gravity dam on the Narmada River near Navagam, Gujarat in India. It is the largest dam. The project took form in 1979 as part of a development scheme to increase irrigation and produce hydroelectricity. The World Bank backed the scheme early in the planning but withdrew its support in 1994. The entire Narmada River basin (Figure 12.7) would be impacted, along with Gujarat and other states, to which water would be diverted.

The Narmada Valley Development Project, as it is officially labelled, is promoted for the production of hydroelectric power. The major reasons for the project include impounding and distributing water for flood control, irrigation, and drinking water.

Spread across the development would be a large number of dams, small to large, more than 3000 if all are completed, which would displace a million and a half people, who must be settled elsewhere. The Sardar Sarovar Dam in Gujarat, with a design height of 160 m, would flood more than 37,000 ha extending well into Madhya Pradesh, covering forests and farmlands and displacing about 500,000 people.

Despite protests, the Supreme Court gave clearance for the height to be increased to 121.92 m. But it also gave directions to Madhya Pradesh and Maharashtra (the Grievance Redressal Authorities of Gujarat) that

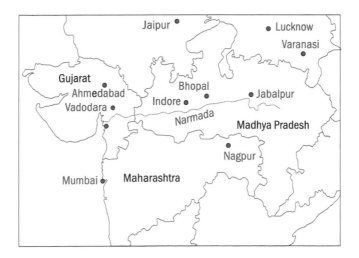

Figure 12.7 Narmada River

before further construction begins, they should certify after inspection that all those displaced by the raise in height of 5 m have already been satisfactorily rehabilitated, and also that suitable vacant land for rehabilitating them is already in the possession of the respective states. This process shall be repeated for every successive 5 m increase in height.

12.7 OIL SPILLS AND IMPACTS ON THE ENVIRONMENT

"If you drill, beware of the spill" could be the slogan of any company drilling for oil. The spill may happen at the well, in transport, or storage. The spills can be small, a few barrels, or can be huge, a large ship. The oil and shipping companies do not want the spills as any accident costs them money. When they lose a ship, losing the cargo costs them, bad publicity follows, and lawsuits are likely.

While oil floats, wind and waves mix it in a complex fashion. Some oil evaporates, but much of the oily sheen stains whatever it touches or forms scattered tar balls. Pictures of oil-coated birds or dead turtles are dramatic, and the volume lost in a major spill is startling. On 28 July 2003, the *Tasman Spirit* ran aground near Karachi, Pakistan, and 12,000 tonnes of oil spilled into the Arabian Sea, contaminating 16 km of the coastline. There have been smaller oil spills in the Bombay High coast of Maharashtra.

An even larger spill than the *Tasman Spirit* and the largest on record for ships is the *Atlantic Empress*. The ship collided with the Aegean Captain, also a tanker, in July 1979 near Trinidad and Tobago. Around 300,000 tonnes were lost from the two ships; fortunately none of the oil reached the shore. Much oil is probably still in the *Empress*, lying in deep water.

Most spills are accidental, from the wrecked tanker to the broken barrel, but some are deliberate. The largest spill was in the Gulf War in 1991. The Iraqi Military discharged more than a thousand tonnes into the Persian Gulf, intending to destroy the water supply of Saudi Arabia and other Gulf countries. Iraq also set the wells in Kuwait on fire, blackening the skies and wasting more oil. On a much smaller scale, a company may secretly discard a few barrels of contaminated oil instead of bearing the expense of proper disposal.

Accidental spills also range from barely noticeable to astonishing. The worst oil spill in Asia was in Uzbekistan. In March 1992, a well blew out in the Mingbulak oilfield and caught fire, burning for 2 months.

Unburnt oil, some 2 million barrels, was also captured in containment ponds. Two million barrels will fill a large oil tanker.

In April 2010 near Louisiana in the United States, a well surged out of control while drilling in the deep waters of the Gulf of Mexico. The Deepwater Horizon, a floating drilling platform, exploded and burnt. The fire continued until the rig sank two days later. The sinking destroyed the connection from the wellhead to the surface and substantial leaking began. As the wellhead was 1500 m down in the ocean, control was difficult. Even knowing what was happening below was difficult.

The amount of crude oil spilled in this incident is probably greater than that from even the biggest tanker accident. Every week, more than 40 tonnes of oil was coming out, the amount carried by a medium-sized tanker. Oil continued gushing out for 12 weeks. As the oil floated, it covered 6500–24,000 km^2, depending on the wind and currents.

BP (formerly British Petroleum) was responsible for the operation. It did not deal well with public perception; at one point, the CEO even said that the spill was not that significant. The facts showed otherwise. BP was not sufficiently prepared to anticipate and prevent the accident. A blowout preventer did not stop the flow. The preventer had not been properly tested, and alarm systems were not operational. Once the accident occurred, BP was not prepared to deal with the emergency. Repeated failures to stop the leak meant that the oil continued to gush. The first attempt to stop the leak was to activate the blowout preventer. That failed. Drilling fluid was pumped into the well to plug it, and a large containment dome was installed. Both failed. Some of the oil that escaped was burnt, some skimmed. Neither method was successful. Finally, after three months, a second containment cap seemed to halt the flow. Two relief wells should divert the flow, at the cost of $100 million each. The wells would reach 5500 m. The operation used ships, underwater vehicles, helicopters, and airplanes at the total expense of several billion US dollars.

To its credit, BP says that it will make good for all the damage done. Regardless, the economic loss is immense; exact calculation is impossible. Not just the valuable oil that was lost, but the fisheries that were closed, contaminated beyond use. Beaches were off limits; even hearing of the spill kept tourists away. Birds and turtles were coated with oil and died.

Petroleum is used because it burns. It is dangerous because it burns and is toxic when it spills. When technology is pushed to the limit—in the Deepwater Horizon case, wells were drilled in deep water—accidents can happen. The results, both anticipated and unexpected, always mean trouble.

References

Brewster Community Solar Garden. 2015. Details available at <www.brewstercommunitysolargarden.com/member-benefits>

Bryan, N. 2004. *Love Canal: Pollution Crisis (Environmental Disasters).* Milwaukee, Wisconsin, USA: World Almanac Library

Choudhury, A., D. K. Lahiri Choudhury, A. Desai, J. W. Duckworth, P. S. Easa, A. J. T. Johnsingh, P. Fernando, S. Hedges, M. Gunawardena, F. Kurt, U. Karanth, A. Lister, V. Menon, H. Riddle, A. Rübel, and E. Wikramanayake. 2008. *Elephas maximus. IUCN Red List of Threatened Species. Version 2010.4.* International Union for Conservation of Nature.

CPCB (Central Pollution Control Board). 2008. *The Scheme: Ecomark.* Details available at www.cpcb.nic.in/scheme_eco.php>

CSE (Centre for Science and Environment). 1999. What goes down must come up. *Down to Earth,* 31 August Details available at <www.rainwaterharvesting.org/Crisis/Groundwater-pollution.htm>

Diamond, J. 2005. *Collapse: How Societies Choose to Fail or Succeed.* New York: Viking Press

Dugger, C. W. 2001. New Delhi Journal; In India's capital, a prayer for the belching buses. *New York Times,* 28 September. Details available at www.nytimes.com/2001/09/28/world/new-delhi-journal-in-india-s-capital-a-prayer-for-the-belching-buses.html?pagewanted=1

Engleson, D. C. and D. H. Yockers. 1994. *A Guide to Curriculum Planning in Environmental Education.* Madison: Wisconsin Department of Public Instruction

Feiden, D. 2012. Super-green restaurant set to open its doors in E. Harlem with 45 environmentally friendly features. *Daily News,* 3 December 2012, <www.nydailynews.com/new-york/uptown/new-appleby-serving-mother-earth-article-1.1212728>

Feldman, D., A. M. Brockway, E. Urlich, and R. Margolis. 2015. Shared solar: Current landscape, market potential, and the impact of federal securities regulation. Details available at <www.nrel.gov/docs/fy15osti/64243.pdf>

Gaston, K. J. and J. I. Spicer. 2004. *Biodiversity: An Introduction*, 2nd edition. New York: Wiley-Blackwell

Global Footprint Network. 2010. Details available at http://www.footprintnetwork.org/en/index.php/GFN/page/footprint_basics_overview

Global Humanitarian Forum. 2009. *Human Impact Report: Climate Change—The Anatomy of a Silent Crisis*. Details available at <www.ghf-ge.org/human-impact-report.pdf>

Guha, R. 2007. *India after Gandhi*. New York: Ecco

IPCC (Intergovernmental Panel on Climate Change). 2007. Fourth Assessment Report Working Group. Details available at <www.ipcc.ch/publications_and_data/publications_and_data_reports.shtml>

IPCC (Intergovernmental Panel on Climate Change). 2014. *Fifth Assessment Report Working Group*. Details available at<www.ipcc.ch/publications_and_data/publications_and_data_reports.shtml>

Kothari, A. and A. Patel. 2006. *Environment and Human Rights*. New Delhi: National Human Rights Commission

Kumari, K. 2007. Evolution of environmental legislation in India. Details available at <http://works.bepress.com/cgi/viewcontent.cgi?article=100 3&context=krishnaareti>

Lal, J. B. 1989. *India's Forests: Myth and Reality*. New Delhi: Natraj Publishers

Lawyer's Collective and the O'Neill Institute for National and Global Health Law. M.C. Mehta vs. Union of India (Kanpur Tanneries), *Global Health and Human Rights Database*. Details available at <www.globalhealthrights.org/asia/m-c-mehta-v-union-of-india-kanpur-tanneries>

National Renewable Energy Laboratory. 2014. *Potential for Shared Solar Projects in the United States*. Boulder, Colorado: NREL

Ramesh, R. 2009. Delhi to Outlaw Plastic Bags. *The Guardian*, 16 January. Details available at <www.guardian.co.uk/world/2009/jan/16/plastic-bags-india-delhi-ban>

Richard, M. G. 2012. Best air filtering plants, according to NASA. *Mother Nature Network*, 30 November. Details available at <www.treehugger.com/files/2009/02/air-filtering-plants-indoors-air-quality-benzen-formaldehyde.php>

Saharia, V. B. (ed.). 1982. *Wildlife in India*. New Delhi: Natraj Publishers

Scott, D. A. 1989. *A Directory of Asian Wetlands*. Gland, Switzerland, and Cambridge, UK: IUCN

Shet, S. 2003. Environmental education finally finds a place in India's school textbooks. *InfoChange India*, August. Details available at <http://infochangeindia.org/20030804258/Education/Features/Environmental-education-finally-finds-a-place-in-India-s-school-textbooks.html>

The Hindu. 2004. Voter turnout: End of the democratic upsurge? *The Hindu*, 26 April. Details available at <www.hinduonnet.com/2004/04/26/stories/2004042603031200.htm>

Times of India. 2009. 57% voter turnout in Phase 4, Bengal tops with 75%. *Times of India*, 8 May Details available at <http://timesofindia.indiatimes.com/india/57-voter-turnout-in-phase-4-Bengal-tops-with-75/articleshow/4495119.cms>

Treehugger. 2007. Is phantom energy haunting your house? *Treehugger*, 12 March. Details available at www.treehugger.com/files/2007/03/is_phantom_ener.php

UNESCO & COMEST. 2005. The Precautionary Principle. Details available at

http://unesdoc.unesco.org/images/0013/001395/139578e.pdf

Wilson, E. O. 1988. *Biodiversity*. Washington DC: National Academy Press

Websites

www.navdanya.org

<www.iso.org/iso/home.html>

www.censusindia.net

www.un.org/popin/wdtrends.htm (a United Nations site)

www.un.org/esa/population/unpop.htm (another United Nations site)

www.prb.org (Population Reference Bureau)

www.census.gov/ipc/www/idb/

www.census.gov/popclock

http://en.wikipedia.org/wiki/Population_pyramid

<www.iccwbo.org/policy/environment/id17226/index.html>

<http://old.cseindia.org/programme/industry/automobile_sector.htm>

www.teriin.org/index.php

www.chintan-india.org

<www.cseindia.org/node/357>

http://edugreen.teri.res.in/explore/laws.htm

<www.cseindia.org/taxonomy/term/20082/menu>

<www.righttowater.info/ways-to-influence/legal-approaches/case-against-coca-cola-kerala-state-india/> and <http://consciouslifenews.com/coke-pepsi-being-pesticides-india/1145862/>

Index

About the Authors

Mr R N Bhargava is Chairman cum Managing Director of Ecomen Laboratories Pvt. Ltd, a Lucknow based company providing expertise in all aspects of Environment and Forestry. He is an executive from Corporate Sector (Tata Steel). He is an Engineer, and MBA from PGDBM (XLRI). Mr Bhargava is also Environmental Auditor from IEMA (also a full member), UK He has 56 years of experience and exclusive 25 years of experience in environmental monitoring, testing, studies and preparation of EIA/EMP reports (nearly 100). Mr Bharagava has authored a few books and contributed articles on environment in journals apart from presenting papers on various aspects of environment in renowned forums.

Dr Rajaram is a geotechnical and environmental engineer with 41 years of work experience in the United States, Canada, and India. At present, he is the Director of Infrastructure for CCJM India, a subsidiary of CCJM in the United States. He is focusing his efforts on Smart Cities proposed all over India. He has a BE in Mining Engineering from Osmania University, Hyderabad, MS from the South Dakota School of Mines and Technology, and PhD from the University of Wisconsin, Madison, Wisconsin, USA. He obtained his degree in environmental and energy law from IIT Kent College of Law in Chicago, Illinois, USA. He has published 3 books, and written over 50 articles in international publications.

Mr Keith Olson has a BS degree from Rocky Mountain College in Montana and a MS from New Mexico Highlands University in USA. Besides teaching, Mr Olson has been active in numerous environmental organizations. In Elmhurst, he was the chairman of a group responsible for a successful community recycling programme. A long-time member of the Sierra Club, he was chairman of the local chapter and the regional outing committee. Since 1977, he has headed groups, school and community, managing a natural area, a local prairie. Since 1996 he has been a member of the county environmental commission.

Lynn Tiede is a National Board Certified, New York City public school teacher with 20 years of experience, teaching social studies, civics, and environmental education. Ms Tiede has worked tirelessly to green her school, local East Harlem community, and NYC. Ms Tiede was employed for two years as the Education Manager for the National Audubon Society. Lynn Tiede received her BA in Anthropology/Sociology from Rhodes College, Memphis, Tennessee MA from the City University of New York. She has worked on various sustainability projects via entities such as Community Board #11, the Manhattan Solid Waste Advisory Board, Borough President Scott Stringer's 'Go Green East Harlem Initiative', and is a two time winner of the Department of Sanitation's Team Up to Clean Up Award for recycling programmes created at her school. Lynn received a Fulbright Indo American Environmental Leadership Fellowship to study environmental sustainability and education initiatives in India in 2005. Lynn has continued to study environmental education and sustainability issues in both China and South Africa through the China Institute (2009) and National Endowment for the Humanities (2012).

Printed and bound by CPI Group (UK) Ltd, Croydon, CR0 4YY

17/10/2024

01775681-0006